SOUTHERN CALIFORNIA EXTENDED

TOURING NORTH AMERICA

SERIES EDITOR
Anthony R. de Souza, *National Geographic Society*

MANAGING EDITOR
Winfield Swanson, *National Geographic Society*

SOUTHERN CALIFORNIA EXTENDED

Las Vegas to San Diego and Los Angeles

BY
LARRY FORD
AND
ERNIE GRIFFIN

RUTGERS UNIVERSITY PRESS • NEW BRUNSWICK, NEW JERSEY

This book is published in cooperation with the 27th International Geographical Congress, which is the sole sponsor of *Touring North America.* The book has been brought to publication with the generous assistance of a grant from the National Science Foundation/Education and Human Resources, Washington, D.C.

Rutgers University Press
109 Church Street
New Brunswick, New Jersey 08901

The paper used in this book meets the minimum requirements of American National Standard for Information Sciences—Permanence of Paper for Printed Library Materials, ANSI Z39.48-1984.

Library of Congress Cataloging-in-Publication Data

Ford, Larry.
    Southern California extended: Las Vegas to San Diego and Los Angeles
    Larry Ford & Ernie Griffin.—1st ed.
      p.   cm.—(Touring North America)
    Includes bibliographical references and index.
    ISBN 0-8135-1884-9 (cloth: alk. paper)—ISBN 0-8135-1885-7 (paper: alk. paper)
    1. Califonia, Southern—Tours. 2. Nevada—Tours. 3. Arizona—Tours. 4.
    California, Southern—Geography. 5. Nevada—Geography. 6. Arizona—
    Geography. I. Griffin, Ernie. II. Title. III. Series.
F867.F673   1992
917.9—dc20
                                    92-10530
                                    CIP

First Edition

Frontispiece: Spanish baroque architecture, Balboa Park, San Diego.

Series design by John Romer

Typeset by Peter Strupp/Princeton Editorial Associates

△ **Contents**

△ Foreword

Touring North America is a series of field guides by leading professional authorities under the auspices of the 1992 International Geographical Congress. These meetings of the International Geographical Union (IGU) have convened every four years for over a century. Field guides of the IGU have become established as significant scholarly contributions to the literature of field analysis. Their significance is that they relate field facts to conceptual frameworks.

Unlike the last Congress in the United States in 1952, which had only four field seminars, the 1992 IGC entails 13 field guides ranging from the low latitudes of the Caribbean to the polar regions of Canada, and from the prehistoric relics of pre-Columbian Mexico to the contemporary megalopolitan eastern United States. Our series also continues the tradition of a transcontinental traverse from the nation's capital to the California coast.

This field examination of Southern California, one of America's most distinctive and controversial regions, analyzes the diverse problems of water supply, urban sprawl, immigration issues, agricultural displacement, industrial expansion, research and development, international commerce, border control, winter tourism, and environmental change. Southern California today, in the American mind, is the prototype of the nation's life tomorrow.

Ernst C. Griffin, professor and chairman of the Department of Geography at San Diego State University, has contributed significant insights to our understanding of cultural and settlement patterns of the Americas. His university colleague, Larry R. Ford, is noted for his acclaimed work on cultural geography and urban design.

Anthony R. de Souza
BETHESDA, MARYLAND

△ Acknowledgments

We acknowledge the dedicated work of the following cartographic interns at the National Geographic Society, who were responsible for producing the maps that appear in this book: Nikolas H. Huffman, cartographic designer for the 27th IGC; Patrick Gaul, GIS specialist at COMSIS in Sacramento, California: Scott Oglesby, who was reponsible for the relief artwork; Lynda Barker; Michael B. Shirreffs; and Alisa Pengue Solomon. Assistance was provided by the staff at the National Geographic Society Map Library and Book Collection, the Cartographic Division, Computer Applications, Illustrations Library, and Typographic Services. Special thanks go to Susie Friedman of Computer Applications for procuring the hardware needed to complete this project on schedule.

We also thank Lynda Sterling, public relations manager and executive assistant to Anthony R. de Souza, the series editor; Richard Walker for proofreading the volume; and Tod Sukontarak for indexing the volume and for photo research. They were major players behind the scenes. Many thanks, also, to all those at Rutgers University Press who had a hand in the making of this book— especially Kenneth Arnold, Karen Reeds, Marilyn Campbell, and Barbara Kopel.

Errors of fact, omission, or interpretation are entirely our responsibility; the opinions and interpretations are not necessarily those of the 27th International Geographical Congress, which is the sponsor of this field guide and the *Touring North America* series.

PART ONE

Introduction to the Region

Southern California Extended

△ Southern California

Southern California has captured the American imagination and psyche in a manner unparalleled by any region in the nation's recent history. A unique set of physical features and cultural elements, highlighted by a benign climate and a unique lifestyle, has brought about a perception of a place that transcends reality. This land of beaches, sunshine, palm trees, and "beautiful people" has come to epitomize the current American dream, despite an array of problems ranging from air pollution and rapid population growth to inadequate water supply and earthquakes. America's infatuation with Southern California seems to have grown progressively stronger with each passing decade of this century. Not only do people flock to the region but Southern California's fads and fashions increasingly influence the entire nation. Ironically, while it may be true that "the West is the most American part of America," Southern California is arguably the least typical part of America. Is this what makes our self-styled "Lotus Land" so alluring?

Just as it is difficult to define the exact dimensions of its appeal, there is no universally accepted definition of what the term Southern California means geographically or even if it should be a proper noun. Many people outside the region vaguely envision Southern California as being the Los Angeles/San Diego conurbation highlighted by beach communities such as La Jolla, Venice, and Malibu, with the glitz of Hollywood and Disneyland thrown in. Lots of Southern Californians consider their realm to be something called the South Coast or the Southland, roughly the coastal lowlands of Santa Barbara, Los Angeles, Orange County, and San Diego along with their immediately adjacent interior valleys. Many Southern California surfers do not acknowledge that there is intelligent life east of Interstate 5. In general, geographers define the physical region of Southern California as everything south of the Tehachapi Mountains and west of the southeastern California deserts, an area of about 12,000 square miles. From the perspective of its cultural geography, we like this definition the best. But the term Southern California is also commonly used to describe a much larger area: all of California south of the Tehachapis, including the Mojave and Colorado deserts. This generic region is often considered the counterpart of Northern California for comparative purposes. For the needs of this guidebook, we will use this latter definition of Southern California because the interior areas are functionally related to coastal Southern California and it conveniently suits our requirements.

One thing is certain about Southern California: It is unique physically and culturally. Despite its current attractiveness to so many people, here was an area that remained sparsely populated and virtually unwanted until the mid-1800s. This was due equally to its isolation from major nodes of population and the harshness of its physical environment. Even Southern California's original inhabitants had to struggle mightily to maintain themselves.

THE FIRST AMERICANS AND THEIR ENVIRONMENT

The initial population groups to occupy this region, according to present knowledge, were part of the migrations of Asians into North America which occurred during the final glacial period of the Pleistocene, some 20,000 to 50,000 years ago. Anthropologists and others who study the origins of Americans differ on precisely when these folks began filtering onto the continent, but opinions range from 10,000 and 40,000 years before the present (BP). It is thought that small groups of semi-nomadic hunters walked across what is now the Bering Strait via a land bridge created by the decline of sea level resulting from the massive quantities of water held in continental glaciers. Others argue that glacial ice linked the continents and was traversed by these ancestors of American Indian peoples. Similar agreement is lacking on when the earliest groups established themselves in Southern California, but most estimates place the time-frame between 4,000 and 10,000 years ago.

Southern California's Indian populations belong to two major families, Shoshones and Hokans. However, linguistic variety was tremendous among the numerous small, subtribal groups of people. Of the 150,000 or 200,000 American Indians estimated to have been in California at the time of European arrival, less than a third of them lived in the south. The region's Indians were technologically of the Stone Age and lived in relative balance with their environment. Unlike their more sophisticated contemporaries in Mesoamerica, these people left few cultural artifacts and little imprint on the landscape for us to recall their presence. Except for the Yumas and Mojaves, primitive agriculturalists who occupied the floodplains of the Colorado River, all of California's indigenous peoples were hunters and gatherers. These population concentrations were limited in size, usually having fewer than 150 persons who occupied territories that averaged perhaps a 100 square miles in area. They survived by gathering acorns, which they ground into meal, and other plants, and by hunting small game and marine life, especially fish and clams.

The first reality to be grasped in respect to Southern California is that it is dominated by an arid and semiarid climate. Except in the higher elevations of the various local mountain systems, few places in the region receive more than sixteen inches of precipitation annually. For example, San Diego averages less than ten inches of rainfall a year, Los Angeles about fourteen, and Santa Barbara around sixteen. The interior deserts lie in the rain shadow of the Peninsular Ranges and receive less than three inches of precipitation a year. The good part of this situation is that the sun shines a lot in Southern California, in most places more than 340 days a year. The flip side is local water is scarce except in rainy periods. In the dry summer months even many of the major streams dry up. Hot summer and mild winter temperatures produce high levels of potential evapotranspiration.

In the low-lying coastal basins this semiaridity is clearly reflected in the natural vegetation assemblages, where coastal sage and chaparral formations predominate. Coastal sage is primarily restricted to the coastal hillsides and consists of evergreen shrubs, such as California sagebrush, blue sage, and white sage, along with wild buckwheat. Usually three to five feet in height and forming a fairly open landscape, coastal sage vegetation is especially pretty when in flower at the end of the winter rains. By contrast, chaparral vegetation is much more widespread throughout the region. Chaparral is comprised of evergreen shrubs which come in a variety of species and grow in dense stands from six feet to ten feet high. Generally the leaves are small, hard, and sometimes drought deciduous. Such plants as chamise, ceanothus, manzanita, and scrub oak are commonly associated with chaparral. The pungent aroma of chaparral on a hot summer day is one of the unforgettable sensory experiences of Southern California. Chaparral also burns easily when dry, which facilitates the great wildfires peculiar to this region, especially during periods of dry "Santa Ana" winds, strong, dry, easterly winds accompanied by abnormally high temperatures and very low humidity. Before the arrival of Europeans, virtually the only trees in the coastal lowlands of Southern California were found along the riverine valley traces. Above the 3,000-foot level in the area's many mountain ranges, assemblages of oak woodlands, piñon pine/juniper forests, and majestic pine forests occur.

Eastward beyond the mountains, the Mojave Desert's dry surfaces are dominated by countless widely spaced creosote bushes interspersed with clumps of grayish burroweed. Pungent in aroma and dirty green or olive in color, the creosote bush grows to a height of about three feet. Mesquite, ocotillo, yucca, and an occasional palo verde are interspersed with creosote bushes throughout the Mojave while California golden poppies, the state flower, and other annuals carpet the desert floor after wet spring months. In the hotter, drier Colorado Desert to the south, creosote bushes and burroweed remain the most common plants but are much more widely spaced than in the Mojave. Cacti, especially the cholla, and ocotillo are also conspicuous.

In a nutshell, this region was hardly paradise on Earth. This combination of climate and vegetation ensured that the groups of native Americans who lived in various parts of Southern California, technologically limited as they were, were few in number and, during drier years, were hard pressed to survive.

THE SPANIARDS

California was the name the Spanish writer, Garci Ordonez de Montalvo, gave to a mythical island he described as being located "on the right hand of the Indies . . . very close to Earthly Paradise." What prescience! Not even the Golden State's Chamber of Commerce could top that. Hernan Cortéz, conquistador of Mexico, is credited with giving the name California to all of the Spanish claims west of the Sea of Cortez (the Gulf of California). Spain evinced very little interest in what is now California, or any of its lands along the West Coast, until the second half of the 1700s. The great deserts to the east and the treacherous oceans to the west effectively isolated California from the rest of Spain's realm in the New World. Although the Manila galleons used the prevailing winds and currents on their return voyages from Asia and arrived off the coast of California prior to sailing southward to Mexico, they made infrequent landfalls and then often with devastating results. Only the 1542–1543 voyage of João Rodriguez Cabrilho (Juan Cabrillo)—a Portuguese captain in the employ of Spain who made it as far north as Monterey Bay—and the 1602 expedition of Sebastian Vizcaino, who named San Diego Bay and most of the coastal inlets and headlands, touched Southern California before 1769.

Nearly two and a half centuries elapsed from its discovery until the Spanish even tried to settle the region. By the 1760s the Russians, who were interested in Alaska, had founded outposts as far south as the San Francisco Bay area and the British were actively interested in colonizing the Oregon Territories. The fear of losing claims to these countries, especially to their English rivals, prompted King Carlos III to establish Spanish hegemony in California. He named Gaspar de Portola, now virtually forgotten for his role, as commander of a land and sea expedition to colonize the region. Junipero Serra, the Franciscan priest who accompanied the overland expedition that established the mission system from San Diego to San Francisco, has come to personify the Spanish settlement of California.

The Spanish used a tripartite colonization plan of presidios, missions, and pueblos. The presidios were military outposts set up to defend strategic locations. Of the four presidios built, San Diego and Santa Barbara were in Southern California. The Spanish erected twenty missions, including four at presidio sites. Two pueblos (towns), Los Angeles and San José, were founded as a part of the colonial design but a number of others grew up around presidios and missions. All of these settlements had to be provisioned by ship until about 1780 when the overland route Juan Bautista de Anza had forged across the deserts (1774 to 1775) was put into use. As a result, virtually all settlements were located near the coast. Agricultural production from mission lands provided much of the food needs for the population while hides from cattle introduced by the Spaniards formed the basis of the commercial economy.

Spain's reign in California lasted slightly more than fifty years, from 1769 to 1821, and its impact was surprisingly limited. As elsewhere in Spain's New World colonies, a single port had the legal right to all imports and exports. Thus, Monterrey was the capital and economic heart of California. But because of the restricted economic opportunities and its isolation, few Spaniards were attracted to the region. To put this into perspective, in 1820 there were only slightly more than 3,700 Europeans in all of California, the majority in the southern coastal areas. At its height, the Spanish missionary efforts yielded a little over 50,000 converts from the Indian population. Furthermore, fewer than twenty-five land grants were made by the Crown in California and most of these were in San Diego and Los Angeles. Aside from establishing the initial settlement patterns and names for cities, perhaps the most enduring contribution of the Spanish era was the architectural style introduced in the missions, with their graceful arches and tiled roofs. In Southern California this became the basis for the "Mission Revival" building styles that epitomize the area.

THE MEXICAN PERIOD

Mexico won its independence from Spain in 1822 and claimed as its own all of the lands of New Spain formerly held by the Crown. Until the Mexican–American War of 1848, Mexico administered California from the capital of Monterrey (in what is now Nuevo León). While facing many of the same problems of inaccessibility and limited opportunity that had plagued the Spanish, population in the area grew substantially under the Mexicans. By the end of their short rule nearly 7,000 Mexicans and 1,000 other Europeans lived in California, an increase of more than 100 percent in twenty years. Much of this growth was spurred by Mexico's more generous policy of awarding land grants (haciendas), which opened up broader opportunities for new settlers. Nearly 800 haciendas were issued during this period. Simultaneously, however, the legal rights of the Church were severely limited by the new Mexican government. Mission lands and cattle were confiscated and the missions were left with only their function as parish churches. The effect on the missions' Indian populations was dramatic: It declined by nearly two thirds in less than a decade. Many of the people became workers on the newly formed Mexican "ranchos."

Meanwhile, several hundred "foreigners," mostly from the United States, established themselves in California during this time. Because the economy depended overwhelmingly on the production of cattle, Mexico's removal of trade restrictions imposed by the Spanish prompted a thriving, albeit long-distance, commercial interchange to develop with the United States. Hides and tallow were tremendously abundant, especially in Southern California, and these products were exchanged for manufactured goods produced in newly industrializing New England. Southern California played an important role in this trade as San Diego became the primary collection center for hides brought from the grazing lands of Santa Barbara, the Los Angeles basins, and the interior valleys such as San Bernardino. Boston merchants sent fast clipper ships around Cape Horn to California. (William Henry Dana's classic, *Two Years before the Mast,* vividly recounts this.) U.S. merchants, most of them New Englanders, settled in California, gaining a strong influence in the burgeoning trade as both buyers and sellers as well as bankers by providing credit to the Mexican ranchers. A number of U.S. citizens also acquired tracts of land from the Mexicans and entered into the cattle-ranching business.

In addition, whalers plied the coast of Southern California taking huge gray and sperm whales for oil and buying provisions from locals. In an interesting twist, the cattle introduced into the Hawaiian Islands ranches were imported from California during the Mexican period, complete with *vaqueros* (Mexican word for cowboy) to teach ranching techniques.

These commercial links between California and New England would increase in importance after the area was annexed to the United States. Many of these early entrepreneurs and landowners became vital cogs in the Americanization of California. The romantic Zorro-like image from the region's short-lived Mexican period has been pervasive, but despite its storied ties to Spain and Mexico, in reality California had only a thin veneer of Hispanic culture. Numerically, culturally, and economically, the region's Hispanic roots were relatively shallow. After eight years of colonization, only about 7,000 people of Spanish-speaking origin occupied (Alta) California (all of California above Baja California) and much of the growth that occurred under the Mexicans was concentrated in Southern California. It was the region from Santa Barbara to San Diego that became the most prosperous in Alta California. However, that would change very quickly.

ANNEXATION BY THE UNITED STATES

California was taken away from Mexico in the summer of 1846 when the United States' Pacific Squadron, a naval force stationed in Mazatlan and headed by Commodore John Sloat, and the land forces of Captain John Frémont's "California Brigade" joined forces and occupied Monterey. The only real battles associated with this U.S. conquest occurred in Southern California, at Los Angeles and San Pasqual, northeast of San Diego. California was administered by a military government until it was admitted to the Union in 1849. Ultimately, under the terms of the Treaty of Guadalupe Hidalgo, the United States would pay Mexico $15 million for the lands of California, Texas, New Mexico, and Arizona.

In reality, the United States had very little interest in Southern California. The prize was San Francisco Bay, which eastern speculators equated with California. They envisioned the great valleys of northern and central California as an alternative to Oregon to waves of westward settlers. San Francisco Bay would allow easy ocean access which would benefit East Coast interests, especially those of New England who had already established ties with the region. By contrast, Southern California was of little import because its dry climate limited agricultural opportunities and, therefore, large-scale settlement. With the discovery of gold and the onset of the Great California Gold Rush, the southern part of the state became a backwater in relation to the areas of frenetically rapid growth to the north. The number of people in California (excluding American Indians) zoomed from around 8,000 in 1848 to nearly 225,000 in 1852! The overwhelming mass of these people, who arrived by sea and overland, were attracted to the "Mother Lode" country in the foothills of the Sierra Nevada.

For the first several years after the onset of "gold fever," the large rancheros of Southern California prospered as never before because they trailed their large herds of cattle to the burgeoning markets in central and northern California and sold them for astronomically high prices. However, by the late 1850s new herds of animals had been established in the Central Valley and elsewhere nearer to the mines and emerging towns of the gold region. Southern California experienced a severe economic downturn as even the long-dependable linch pin of the culture—livestock ranching—faltered. Not until the 1880s did Southern California really begin to come into its own.

THE BOOM OF THE 1880S

Sectional differences between the north and south of California have been marked since the onset of Spanish occupation. With the capital in Monterrey, the north held unquestioned political and economic sway over Alta California. With the advent of Mexican rule, Southern California became by far the most prosperous and economically important part of the region, but political power was still centered in Monterrey. This was a decided disadvantage for, and irritation to, Southern Californians. The Gold Rush, the emergence of San Francisco as the central place of the entire West Coast, and the city's emergence as the terminus for the newly completed transcontinental railway system gave Northern California the needed impetus. Southern California languished culturally and economically while its population stewed enviously as progress to the north continued apace.

To make matters worse, a series of natural calamities struck Southern California, ultimately ruining the old ranchero system and the culture it had spawned. A severe drought in 1856, unprecedented flooding in 1861, the prolonged drought of 1862 to 1865, and, as if this were not enough, the great grasshopper plague of 1863, virtually wiped out cattle herds and bankrupted many landowners. Nonetheless, the demise of the huge rancho landholdings and the decimation of the great cattle herds of the region created unexpected benefits in that land became available for new settlers and new economic innovations. As a result of the break-up of the massive haciendas, a wave of new U.S. homesteaders arrived in Southern California and they brought with them new agricultural ideas. Thus, in the 1860s and 1870s small-scale farmers experimented with a wide variety of crops, such as grapes and citrus, that would become the hallmark of the region. Along with the collapse of the Gold Rush in the north, Southern California was ripe for growth.

The 1880s produced the first of a series of enormous boom periods for Southern California which have continued with only minor interruptions to the present. Most of the growth during the fabled boom of the 1880s was concentrated in the Los Angeles Basin as a result of its selection as the terminus for the southern transcontinental rail link. Los Angeles was initially tied to a vast new hinterland in the Southwest when the Southern Pacific Railroad route was established through the Central Valley. In the mid-1880s, the Santa Fé Railroad completed its southern transcontinental link and it ended in Los Angeles. This set off a series of events, including the widespread introduction of citrus crops, massive real estate speculation, and winter tourism, which were to change Southern California forever. Orange and lemon groves, fruits prized in eastern markets, spread through the L.A. Basin in the 1880s because they were nicely suited to the semiarid climate, and refrigerated (ice) freight cars allowed them to be moved out of the region efficiently and economically. Large subdivisions were begun and Southern California was advertised throughout the East and Middle West as an American Mediterranean and winter delight. Additionally, Southern California quickly became acclaimed for its salubrious climate, which led to the establishment of towns such as Pasadena by wealthy eastern health addicts. The image of owning a small citrus ranch in a warm and healthful environment proved very appealing as population exploded. The city of Los Angeles, far and away the most important town in the Basin, grew from a couple of thousand in 1880 to more than 50,000 residents by 1890, and exceeded 100,000 in 1900. Coastal towns like Santa Monica and Long Beach as well as places in the San Fernando, San Gabriel, and San Bernardino valleys also sprang to life during this era. Oil was discovered in the Basin in the 1890s and would help to fuel the "Second Boom" of the early 1900s.

Farther south, San Diego also experienced significant growth. The region had been eclipsed by the emergence of Los Angeles, despite its excellent natural harbor. Connected only by a railway spur from the north, San Diego was destined to grow up in the shadow of its more prosperous and dynamic neighbor. Then gold and silver discoveries in the mountains east of the coast led to a minor gold rush and San Diego's population spurted. The entire county had almost 35,000 people in 1890, more than four times higher than a decade earlier, and the city of San Diego grew to over 16,000 residents. Land speculation in the 1880s led to the movement of the downtown from its traditional site on the San Diego River, so-called "Old Town," to its present location on the harbor. The same advertising ploys used to lure folks to L.A. were tried for San Diego but they met with less success. In fact, San Diego stagnated during the 1890s and ended the century with virtually the same population as it had had ten years earlier.

THE SECOND BOOM

Although the 1880s and 1890s saw the first explosion of Southern California's population, that was really a small "pop" compared with what was to come. Nearly 99 percent of this region's population growth has occurred since 1900! Between 1902 and 1914 Southern California, and especially communities in the L.A. Basin, experienced a second major boom period. The oil industry, which initiated production at the end of the century, became a major source of growth. Some of the largest fields brought into production were near the coast which led to the construction of massive refineries and the easy exportation of petroleum products from these facilities. The capital generated from oil spurred a boom in office construction in downtown L.A. as well as the development of a number of blue-collar communities which grew up around the oil fields to accommodate workers. Ironically, the widespread availability of fuel prompted local manufacturing development, and the start of an automobile-oriented society that would make Southern California infamous.

While citrus production continued to prosper and expand, especially southward into its namesake, Orange County, the rise of the movie industry here further stimulated the economy. Attracted by the sunny year-round climate and the varied landscapes accessible within short distances, the Los Angeles metropolitan area, notably Hollywood, became the film capital of the world. Numerous towns and cities sprang up in the Basin during this time, to become the basis for the urban mosaic of the current conurbation. The communities, dependent solely on local wells and streams for their water supply, completed the Los Angeles Aqueduct in 1913, which provided virtually unlimited water to the area. Moreover, Los Angeles built an artificial harbor in San Pedro–Wilmington to replace the old ocean piers it had used previously. The Basin was well-placed to take advantage of the opportunities presented by the opening of the Panama Canal. By the end of this period, the metropolitan area had more than 750,000 inhabitants and, by 1920, Los Angeles, with some 600,000 inhabitants, was among the ten largest cities in the country.

During this period San Diego prospered by becoming a "Navy town." The United States wanted to become a "two ocean" international power, and San Diego offered free land and other inducements to attract a large contingent of naval ships and facilities. The great 1914 California Exposition staged in San Diego to celebrate the opening of the Panama Canal (and to piggy-back on the interest created by San Francisco's World Fair), created Balboa Park, a 4,500-acre urban park in what has become the core of the city. This area's population grew to more than 100,000 by 1920 and about 75,000 of those inhabitants lived in the city of San Diego.

To the north, the idyllic coastal community of Santa Barbara began to attract monied easterners who built mock-Spanish houses on the hills overlooking the ocean. Destined to become one of the state's most affluent and attractive communities, Santa Barbara's relative isolation has kept it an upper-income residential town for several decades. Agricultural production also spilled over the Transverse Ranges into the Simi Valley. Throughout the region the citrus groves and truck farms attracted the large influx of farm laborers from Mexico who formed the basis for the enormous Hispanic community that has emerged in the region. Despite impressions to the contrary, few of these people had previous ties to California.

Southern California's desert communities began to develop during this period also. The first irrigation projects in the Imperial Valley opened up what has become one of the United States' most productive agricultural regions. Dates and grapes were introduced into the Coahuilla Valley at this same time. In 1905, the Colorado River flooded into the newly formed irrigation canals of the Imperial Valley, and the Salton Sea, a 40-mile long lake, was formed in the lowest portion of the Salton Trough.

During the Second Boom a Southern California landscape and lifestyle was beginning to take form. Palm-lined streets became commonplace in nearly all of the region's towns and cities while single-family "bungalows" on large lots became the preferred house type. Easy access to beaches and mountains, combined with a balmy climate, led to an easy-going, outdoor way of living that has evolved into a Southern California hallmark.

THE BOOM OF THE 1920S

World War I caused a brief respite in the pace of growth begun in 1914, but Southern California boomed again during the 1920s. The Los Angeles Basin became the fastest growing part of the nation. To the south, San Diego also changed but at a somewhat slower pace, while the rest of Southern California experienced modest change.

The Los Angeles Basin is composed of a coastal plain and a series of inland valleys separated by hills or low mountains. In turn, the entire Basin is surrounded by several mountain ranges— the San Gabriels, San Bernardinos, and San Jacintos being the most prominent. The coastal plain extends roughly from southeast to northwest while the San Fernando, San Gabriel, and inland valleys to the north have a general east–west orientation. This separation of the Basin into relatively discrete physical units gave people living in the region a series of environmental choices, based on climate and landscape. At the same time, this physiographic pattern, combined with the prevailing westerly winds and a propensity for temperature inversions, created ideal conditions for what was to become known as smog. The mountains on the north, east, and southeast rise to over 10,000 feet in elevation and act as a barrier to air moving out of the Basin while the inversions, which normally have a top level of about 3,000 feet, act as a cap to trap pollutants. Local Indians who lived here before the Europeans did called this area "land of the smokes" because of the lingering smoke plumes that occurred after fires.

Within the L.A. Basin the city of Los Angeles is by far the dominant political and economic entity. However, there are nearly 100 incorporated towns and cities within the Basin. Most of these date from the end of the 1800s and the beginning of the 1900s, but new entities are still being formed. One wag referred to the area as "eighty towns in search of a city." Now, to the uninitiated, the transition from one city to another is not noticeable in the nearly continuous urban agglomeration in much of the Basin. However, this was not always the case.

During the 1920s, Los Angeles economic activities surged, with a concomitant population influx. The oil industry, which began early in the century, exploded as new fields were discovered and brought into production closer to the coast. The oil industry provided a significant portion of the capital that stimulated the region's growth. Simultaneously, earlier population growth coupled with new waves of migrants from the midwest and eastern United States created opportunities for an array of local consumer industries to develop. These, in turn, provided new jobs for more people which created a spiral of growth. Because of the distances between towns and the newness of development in the area, Southern California became the first automobile-oriented region in the world. By the 1920s, L.A. had the highest percentage of auto ownership in the nation, which allowed a number of additional industries— tires and automobile production, for example—to take root. The motion picture industry, begun here only a decade earlier, soared as movies replaced vaudeville. By the end of the 1920s Hollywood had a stranglehold on the movie industry. Finally, the benign climate of the region, which permitted outdoor construction and reduced building costs, led to the development of a large-scale aircraft industry in the Los Angeles Basin and in San Diego. L.A. was destined to become an international leader in the aerospace industry.

Combining these activities with the emergence of the facilities at San Pedro–Wilmington as one of the closest West Coast port facilities to the Panama Canal, excellent rail connections to the East, and the emerging transcontinental highway Route 66, which began in Chicago and ended in Los Angeles, the Basin was truly well-positioned to thrive. The region's hinterland expanded as L.A. became the regional metropolis for much of the southwestern United States and the import–export center for many goods moving to Asia as well as the East Coast.

As a result, Southern California became a land of opportunity blessed with what had become accepted as an ideal climate. By 1930 the city of Los Angeles had well over a million inhabitants and was the fifth largest city in the country. And the smaller towns and cities of the Basin grew even more rapidly than L.A. during this period while the city's share of the total metropolitan population of the Basin began to decline. Many of the bedroom communities closest to Los Angeles began to fill in their urban areas, beginning what would become an enormous conurbation.

Elsewhere in Southern California, the L.A. Basin's phenomenal growth had a spill-over effect. For example, many of the migrants who found their way to San Diego arrived first via Los Angeles. This was in part due to the fact that the sand dunes of the eastern Imperial Valley had to be crossed by a wooden plank highway and this restricted road access to San Diego to only the hardiest souls. Nonetheless, although much smaller in scale, San Diego's population doubled in the 1920s to more than 150,000. The aircraft industry and the growing military installations provided much of the employment for newcomers. Resort communities as diverse as Palm Springs in the desert, Big Bear in the mountains, and beach towns such as Balboa/Newport boomed because of the prosperity of L.A. Meanwhile, agricultural opportunities in surrounding rural areas also soared as truck crops for local consumption and citrus for eastern markets were in great demand. Orange and Riverside counties witnessed massive expansion of their citrus orchards which greatly increased the demand for seasonal labor.

By the end of the 1920s, Los Angeles and its surrounding communities had attracted a large Mexican population, much of which would be drawn into commercial and industrial activities as well as agriculture. Despite the expulsion of many Mexicans during the xenophobia spurred by the Great Depression, the basis for a significant Hispanic community had been laid. Furthermore, the Basin also provided many opportunities for Japanese farmers who migrated to the area in large numbers. Los Angeles's "Japan Town" dates from this era and the Japanese community has long been important there. Other ethnic groups, especially the Chinese, were attracted to the region in the 1920s and, along with Mexican and Japanese, formed the cornerstone for the multiethnic, multiracial, polyglot society which has emerged in the greater Los Angeles metropolitan area. Today more than 213 foreign languages are spoken in L.A.

The Great Depression of the 1930s slowed the unprecedented growth of Southern California but did not stop it completely. Many of the so-called Okies and Arkies who fled the ravages of the Dust Bowl and the collapse of small-scale agriculture in the Middle West and South found their way to the region. Nonetheless, the Boom of the 1920s ended soon after the Great Crash. Then in the late 1930s, Boulder Dam became a tourist attraction, and at the same time, Mexico outlawed gambling, so Las Vegas became a tourist attraction, especially for Southern Californians.

PART TWO

The Itinerary

 Prologue to Paradise

The greater Southern California area, with its extension into Nevada and Arizona, constitutes a remarkably diverse physical and cultural landscape. Physically, you can visit the highest point in the continental United States (Mount Whitney, 14,369 feet) or the lowest (Death Valley, –284 feet), balmy coastal beaches with exotic palm trees or barren interior deserts covered with anything from dunes to Joshua trees. Culturally and socially it boasts everything from one of the world's largest metropolitan areas in Los Angeles, to small agricultural towns such as Westmorland dominated by recent Mexican immigrants, from the elite golf-course–studded playgrounds of Palm Springs/Palm Desert, to the gambling center of Las Vegas.

For two and a half days we will travel through the "empty quarter" of the American Southwest. Although we will stop at the growing metropolitan area of Las Vegas, most of the journey will be through uninhabited or sparsely inhabited desert. The second half of our field experience, on the other hand, will be through what is rapidly becoming one of the largest and most densely populated centers in the world. Our San Diego–Orange County–Los Angeles County route will take us through the home of roughly 14 million people with another 5 million living in nearby cities that we will not visit, such as Tijuana, Mexico, and Santa Barbara, California. It is, as they say, a land of contrasts. Sit back and enjoy the ride.

Kingman, Arizona, to Las Vegas, Nevada

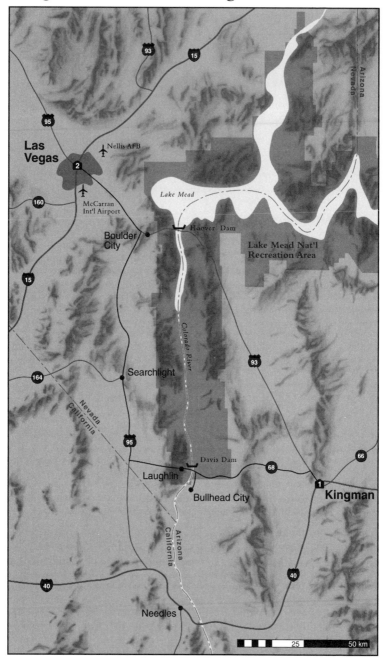

△ *Day One*

KINGMAN, ARIZONA, TO LAS VEGAS, NEVADA, 134 MILES

Kingman, Arizona, to Boulder City, Nevada, 111 miles, Hwy. U.S. 95

We begin exploring this region by leaving Kingman and descending precipitously some 3,000 feet from the Colorado Plateau down to the Colorado River valley. Our first stop will be the greater Bullhead City/Laughlin metropolitan area, newly developing haven of gambling and vice. Entering *Laughlin* you see a miniaturized reproduction of Las Vegas, including oversized paddle-wheeler casinos, high-rise luxury hotels, and monstrous, gaudy neon signs. The reasons for this phoenix rising from the riverbanks in the middle of nowhere are multiple and have to do with site and situation. The site of this new development is directly below the recently constructed Davis Dam, which created a recreational lake and opportunities for boaters, water skiers, and fisherman as well as low-cost energy supplies. From a situational perspective, Laughlin/Bullhead City lies on the southern edge of the Lake Mead National Recreational Area, accessible to folks visiting here as well as for residents from Arizona and Southern California. Laughlin provides a small bit of the glitter and excitement of Las Vegas while simultaneously offering a low-key environment that appeals to "Snow Birds" (people who go to southern Arizona or southern California to escape winter weather) and the geriatric Winnebago set (large recreational vehicles).

Typical casino strip, Laughlin, Nevada.

Surrounding the casinos of "downtown" Laughlin, a linear development hugging the river, are myriad mid-density condominium complexes and long-term motor-home parks. Many of the condos are time-share facilities aimed at short-term visitors who come primarily in winter, attracted by the Southwest's mild temperatures. By contrast, *Bullhead City,* which grew as the workers' community for the dam project, is dominated by the more traditional residential landscapes of desert oasis towns of this area.

Growth has been spectacular as new capital has been channeled into the river resort. Despite the fact that Laughlin is an emerging architectural enigma and a barrel of laughs, we will need to hit the road if we are to arrive at Las Vegas at a timely hour.

Leaving Laughlin we climb onto the riverine terrace of the Colorado Desert, and in 22 miles encounter Highway 95 and head north toward Boulder City and Hoover Dam. Traveling through the sagebrush- and creosote-bush–covered Paiute Valley, we pass through crossroads towns with intriguing names such as Searchlight while the surrounding volcanic peaks reach heights of 5,000 to 7,000 feet. After nearly 50 miles of hot, dusty, and monotonous travel through sparsely populated desert, we come upon the bright lights and greenery of Boulder City.

Nestled in the mouth of a narrow gorge, *Boulder City* was established as the headquarters and residential center for construction of Boulder Dam during the 1930s. Boulder Dam was renamed Hoover Dam (after President Hoover) in 1941. Looking very much like a misplaced Middlewestern town with its green parks and comfortable, solid housing stock, the town has benefited greatly from the abundant water and energy resources available to it from the dam project. It still serves as the dam's headquarters and provides the only tourist facilities for visitors to Hoover Dam and the Lake Mead recreational campgrounds. The well-maintained government-built headquarters complex on the northeast side of the city is a classic example of Depression Era federal building projects and is reminiscent of some of the best new town developments (utopian schemes) of the 1930s.

Following the sinuous and scenic road extending some 7 miles from Boulder City to Hoover Dam, we descend to the seemingly inaccessible site of the dam itself. Begun in 1931 and completed in 1935, Hoover Dam was one of the engineering marvels of its time. An awe-inspiring accomplishment even today, the dam is, at 726 feet, still among the tallest in the world. It is also big. It is 660 feet thick at the base and 45 feet thick at its 1,244-foot-long, curving top and contains more than 3.25 million cubic yards of concrete. Behind the dam is Lake Mead, a giant reservoir covering 158,000 acres with a depth to 500 feet. From 1939 to 1949 Hoover Dam was the world's largest hydroelectric facility.

Take advantage of the tours that are offered and descend into the bowels of the beast to explore the inner workings and hidden mechanisms of this hydroelectric giant. If time permits, view a film on the history of the construction of the dam and the problems encountered in its creation. Then its on to Las Vegas to see where a large portion of this electricity goes. In less than an hour we will be in "Glitter Gulch."

Boulder City to Las Vegas, Nevada, 23 miles, Hwy. U.S. 95

Of all the major cities in the country, *Las Vegas* is among the newest. Mormons settled here in 1855 because of the site's fertility and the availability of water for agriculture; the Las Vegas mission was intended to bring Mormonism to the Paiute Indians and to serve as a communications link to Mormon settlements in Southern California. The Mormons left in 1857 and the area was incorporated by the railroad company as the Las Vegas Ranch in the 1860s, using some of the structures the Mormons abandoned. At the turn of the century, however, Las Vegas was still nothing more than an extensive ranch. In 1902, the railroad's decision to make Las Vegas a division point on the route connecting Salt Lake City to Los Angeles engendered the city of Las Vegas. By 1905 much of the ranch had been purchased by the San Pedro, Los Angeles, and Salt Lake Railroad and the Las Vegas Land and Water Company. More than 1,000 lots were auctioned in that year, thus beginning the speculative growth of the city. However, for the next twenty-five years Las Vegas was little more than a typical western railroad town, not dissimilar from present-day Barstow, California.

Because of its remote location and rough 'n' ready atmosphere which derived from its being a center for ranching and mining, Las Vegas grew up with a reputation as a hard-drinking, saloon-filled gambling town. The city's founders made a stab at respectability in 1909 by banning gambling. And with the advent of the Volstad Act prohibiting liquor in 1918, Las Vegas had little other than the railroad to sustain it. As late as 1930 Las Vegas was hardly unique. With a population of only 5,000 people and a main street lined with gas stations and small shops, it was not much more than a wide place in the road. Still, Las Vegas did everything it could to play up its "Wild West" image. Town leaders sought to attract tourists and Hollywood filmmakers to its western atmosphere. Forgettable "classics" such as *The Railroad Girl* and *Girl on the Trestle* were filmed here.

The story of modern Las Vegas really begins in the 1930s. Its first inducement for growth actually came in 1928 with the announcement that Boulder Dam would be constructed nearby. Shortly thereafter, an influx of capital and infrastructural development began, paving the way for the beginning of the dam's construction in 1933. After completion in 1938, the dam provided Las Vegas a nearly inexhaustible supply of water for urban and recreational use, such as golf courses, as well as a massive source of low-cost electrical power which eventually made possible the unprecedented illumination of "The Strip." In addition, Boulder Dam became an immediate tourist attraction, drawing thousands of visitors from across the country. Lake Mead also provided numerous opportunities for boating, fishing, and fun in the sun for the increasing number of people traveling by automobiles along the expanding network of highways being paved during the Depression.

It is interesting to note that the federal government's intervention played an important role in the area's growth. Likewise, the State of Nevada's re-legalization of gambling in 1931, in part to entertain the massive numbers of workers employed at Boulder Dam, led to a thriving Las Vegas economy even in the throes of the Depression. The heart of the city was Fremont Street in the center of the still compact downtown. In 1932 Las Vegas got its first three-story building, complete with elevator.

By 1940 the city was ripe for a change. Las Vegas entered its golden era in 1941 with the opening of the town's first full-fledged resort, El Rancho Vegas. Unlike the earlier gaming halls and bars, El Rancho Vegas was built in a remote location southwest of the city along Highway 91 leading to Los Angeles. The resort obviously catered to the automobile trade and had as its goal the accommodation of Southern California tourism. Although small by present-day standards and architecturally mundane, El Rancho Vegas was a harbinger of things to come . A second resort on the soon-to-be-famous Strip, the Last Frontier Hotel, opened in 1942. Its slogan, "The early West in modern Splendor," demonstrated

Las Vegas schizophrenia in wanting to present itself simultaneously as a wild-west frontier town and an innovative, luxurious entertainment center. World War II, however, greatly impinged on further tourist development.

During the war Las Vegas's economy was buoyed by the establishment of a military base, currently known as Nellis Air Force Base, and a new magnesium plant. These represented the first major local employers other than the railroad and gambling. The entertainment industry was kept afloat by the few people who could afford or obtain the gas-rationing coupons needed to get there. After the war, however, Las Vegas really exploded!

Southern California, which benefited greatly by wartime industrialization and military growth, had greatly increased its population. By the late 1940s, with a booming economy, plenty of gasoline, and relatively easy highway access, Las Vegas became a tourist destination for the masses as well as the Hollywood avant garde. In 1948 McCarran Field opened as the city's municipal airport and during its first year of operation was served by four airlines and had 35,000 passengers. Although Nellis Air Force Base handled some commercial flights during the 1930s, not until the late 1940s did Las Vegas become accessible by air. Accessibility gave Las Vegas the potential to be a national as well as a regional gaming playground. It also facilitated the movement of highly paid performers who became the star attractions of many of the casinos. Their immense salaries were underwritten by profits from the gaming tables, making it difficult for other venues to compete for their services. Now McCarran Field is one of the most trafficked airports in the western United States. Air transportation has also facilitated other forms of economic activity by making Las Vegas easily accessible to all parts of the country.

The image of Las Vegas as an exotic and slightly roguish place was enhanced by the influx of capital from the criminal underworld and the visible involvement of such characters as Bugsy Siegel in the casino business. In spite of war-related shortages of building materials and labor, Siegel completed construction of the "fabulous" Flamingo Hotel and Casino in record time. Unfortunately for him, Bugsy was rubbed out shortly after the Flamingo's

The Excalibur Hotel, Las Vegas, Nevada.

opening in 1946. In 1950 the Desert Inn, the fifth major resort to open along the Strip, set a new standard for size and luxury. With 300 rooms and its own tennis courts and golf courses, it changed the scale of Las Vegas resorts drastically. The Las Vegas landscape began to offer visual excitement not found elsewhere.

To further enliven the environment, huge oversized neon signs became the trademark of the city. Las Vegas became the symbol for an entirely new kind of American architecture—one meant to be viewed from a speeding car. Casinos began as rather traditional-looking hotels with colorful signs and signature logos, such as windmills or thunderbirds. Gradually the neon signs increased in size and luminosity until they dominated everything around them. During the 1960s author Tom Wolfe noted that the Las Vegas skyline was dominated by signs rather than buildings. By the early

1970s a new phenomenon emerged: The casino/hotels became massive logos themselves. The Strip is now filled with outlandishly oversized replicas of circus tents, Roman palaces, Mississippi stern-wheelers, medieval castles, and Chinese temples. And we are talking BIG: hotels with well over 3,000 rooms and thousands of one-armed bandits (slot machines). Because of this constant search for profit-enhancing novelty, the economic life of hotel/casinos in Las Vegas is quite short. Without constant renovation and enlargement, casinos quickly lose their allure. All of this is meant to create a kind of fantasy world where people can completely forget reality while losing everything from their life's savings to their marital status.

Aside from Disneyland, Las Vegas's architecture is unquestionably the most inventive and spectacular found anywhere in the United States. From a design standpoint, Las Vegas has been, and continues to be, a source of great controversy. While some cringe and blanch at what they consider to be bad taste on an unprecedented and colossal scale, others argue that the city represents innovative and appropriate forms and symbols for the twenty-first century. You can decide for yourself as you cruise among these colorful creations.

With the advent of additional massive resort casinos on the Strip, Las Vegas was able to support a steady flow of major stars who provided almost continuous major-league entertainment. During the late 1940s and early 1950s, the casinos attracted not only major recording stars such as Lena Horne and Frank Sinatra, but also emerging television celebrities such as Milton Berle and Red Skelton. Because of the exceptionally lucrative salaries offered, Las Vegas became one of the prime venues for singers, dancers, and comedians. Elvis Presley and Liberace became indelibly linked to the city's entertainment lore. In fact, Las Vegas began to create its own celebrities, such as Wayne Newton, who became a tourist attraction unto himself. Gaudy and sometimes bawdy revues featuring scantily clad women and lots of feathers, like the Folies Bergère, evolved as a mainstay of Las Vegas entertainment. Later, circus acts, jousts, and magic shows added to the fun.

Beginning in the 1970s and accelerating through the 1980s, Las Vegas diversified. While gambling and tourism remain the major employment and income producers for the city, a wide range of industrial and other activities are growing rapidly. North Las Vegas now boasts a significant base of heavy industry while the metropolitan area generally has benefited from the flight of cost-conscious high-tech and service activities from Southern California. With a population in excess of 600,000 the greater Las Vegas area is beginning to constitute a market for local industries producing a range of consumer goods. Even in economically hard times, Las Vegas's unemployment rates have ranked among the lowest in the nation. Although it remains a classic special-function city, the diversification of its economy is essential if Las Vegas is to avoid potentially cataclysmic problems, such as the legalization of gambling in California. The city's close ties to Southern California facilitate diversification.

While in Las Vegas we will examine some of these phenomena by tracing the evolution of the city from the compact and almost traditional downtown through the various eras of casino construction along the strip. We will also look at some of the industrial zones in north Las Vegas as well as typical residential areas within the urban area. Don't worry, there will be plenty of free time for you to immerse yourself in the debaucheries of your choice during our stay in Las Vegas.

Las Vegas, Nevada, to Palm Springs, California

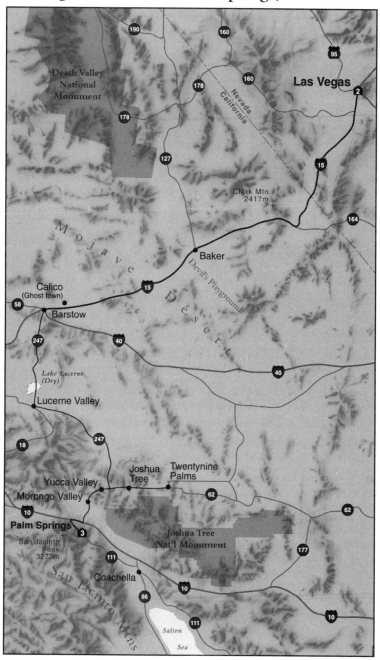

△ Day Two

LAS VEGAS, NEVADA, TO PALM SPRINGS, CALIFORNIA, 253 MILES

Las Vegas, Nevada, to Baker, California, 110 miles, I-15

Talk about monotonous stretches of highway! This must have been why God let the car radio be invented. Leaving Las Vegas at an elevation of 2,033 feet, we ascend southwestward along Interstate 15. For those of you primarily interested in the cultural landscape, we'll wake you up when we get to Barstow. It is about 50 miles to the California–Nevada border through a series of worn mountains and dry lake beds. Physiographically, this area is on the western edge of the Basin and Range Province. The highway passes between the Spring Mountains on the north and the Ivanpah Mountains on the south, crossing Mountain Pass (4,726 feet) shortly after entering California. This portion of the Mojave Desert, which receives less than three inches of precipitation annually, is dominated by a sparse vegetation cover in which creosote bushes are interspersed with grayish burroweed. The pungent creosote bush, greenish or olive in color, grows to a height of some three feet. Mesquite, ocotillo, yucca, and an infrequent palo verde are interspersed among the creosote. Especially after spring rains (the "rainy season"), the California golden poppy adds color to this otherwise drab vegetation assemblage. In 1980 nearly a million and a half acres of the desert south of Interstate 15 was designated as the East Mojave National Scenic Area, the first area in the

United States to be so named. The landscape includes numerous mountains, mesas, volcanic spires, and cinder cones in addition to sand dunes. Many dirt roads and trails lead to primitive campgrounds and recreational areas. Petroglyphs more than 10,000 years old have been found in this region.

Situated in a rain shadow, the Mojave is part of the high desert, with searing summer daytime temperatures exceeding 110 degrees Fahrenheit and surprisingly cool nights, commonly in the 60s. During the winter temperatures dip to well below freezing and snowfall is common. High winds contribute to inhospitable conditions as well as increasing evapotranspiration rates. Indeed, the heavy traffic plying the highway between Las Vegas and Southern California is frequently interrupted by dangerously heavy winds that necessitate closing of the route to vehicles with high centers of gravity. A high-speed rail link, which has been under discussion for many years, would alleviate many of the unpleasantries associated with the journey and allow people to arrive at the casinos already intoxicated.

Baker to Barstow, California, 60 miles, Hwy. 127

After experiencing this desert landscape for 110 miles we arrive at what passes for civilization in the Mojave. *Baker*, at 923 feet above sea level, sits on a small rise between two dry lake beds. Formerly an important gas and water stop on the Los Angeles to Las Vegas route, this small berg contains little more than a group of businesses that cater to tired and thirsty highway travelers. State Highway 127 leads to Death Valley, at 282 feet below sea level the lowest point in the continental United States, some 110 miles to the north, as well as to other equally exotic destinations.

The land rises as we continue westward to *Barstow,* which is over 2,170 feet in elevation. While precipitation increases slightly in this direction, the landscape appears only slightly more luxuriant than that to the east. The city of Barstow, with more than 20,000 people, is the central place of the Mojave Desert region. It was founded in 1886 as a major rail junction and marshaling yard

The McDonald's in Barstow, California.

that linked the routes from southern and central California. Today it still functions as a freight classification and diesel repair site for the Santa Fe Railway. In addition to the railroad, the economy is buoyed by a large military population stationed in surrounding marine facilities, and by highway service functions such as motels, gas stations, and restaurants. A newly constructed thermoelectric plant east of the city—sited on a rail route for transportation—has become one of the major landmarks of the metropolis. Barstow's greatest claim to fame is its enormous McDonald's restaurant, housed in a series of abandoned train cars just off of the interstate (south side of the interstate, east of Barstow; exit Highway 247). Stop here so you can experience a truly American ambience, highlighted by the sight of 300-pound women, attired in cotton house dresses and blue hair curlers, dining with tattoo-festooned male bikers in leather tank tops.

CALICO

Nearby is the ghost town of *Calico,* a state park and tourist attraction. This silver boom town was founded in 1881 and for the next fourteen years was a major mining center with a population that peaked at nearly 3,000. When the price of silver plummeted in 1895, the mines closed and Calico "bit the dust." Its remaining buildings are a monument to the desert's frontier heritage.

Barstow's old commercial center along the old highway north of the freeway (Business 15) has been largely bypassed in favor of a newly emerging suburban-style, freeway-oriented business district. We'll cruise the remains of the early desert community and examine some of the architecture associated with life before air conditioning and modern freeways. The only town of any significance between San Bernardino and Las Vegas, Barstow once served a huge hinterland that it now shares with places such as Victor and Lucerne Valley, which are growing thanks to spillover effects from burgeoning metropolitan Los Angeles.

Barstow to Yucca Valley, California, 77 miles, Hwy. 247

From booming Barstow turn south on California 247 and head for the Lucerne Valley, 34 miles away. Despite its proximity to the population centers of Southern California, the Lucerne Valley remains rustic and serene. Beginning in the 1940s a great deal of land speculation occurred in areas on the far eastern periphery of the Los Angeles metropolitan area in anticipation of development spilling over the San Gabriel and San Bernardino mountains into the expensive lands of the high deserts. Areas such as Palmdale and Lancaster in the north and Apple Valley to the east of L.A. were viewed as prime candidates for everything from a new airport to industrial parks and satellite residential communities. Much of this enthusiasm proved premature. Despite the efforts of developers to sell lots and to attract investment, population remained sparse and haphazard. After decades of fruitless anticipation, rapid growth reached *Victorville* and the I-15 corridor ten to fifteen years ago. The combined populations of the two cities of Victorville and Apple Valley grew from roughly 30,000 people in 1980 to more than 86,000 in 1990 with many more living in unincorporated areas. The community of *Apple Valley* is sometimes called the "Palm Springs of the Mojave" and serves as a bedroom community for wealthy Angeleno executives, many of whom commute by private plane to their offices in the city. Although Victorville has grown significantly because of industrial development, George Air Force Base, and low-cost housing tracts, its main claim to fame remains the Roy Rogers–Dale Evans Museum, complete with Roy's stuffed horse Trigger.

By contrast, the Lucerne Valley remains largely agricultural. Relatively well-watered because of runoff from the often snow-covered San Bernardino Mountains immediately to the south, alfalfa and irrigated pastures give a unique oasis-like appearance to the area. Unlike its Swiss namesake, the region remains relatively undiscovered.

Highway 247 turns sharply toward the southeast, hugging the contour of the mountains to the south as soon as you leave Route 18. Over the next 43 miles, the scattered but pervasive groups of shacks on desert lots interspersed with undercapitalized weekend resorts are testimony to the unsuccessful speculation that has drawn people to the area since the 1940s. Not surprisingly, nearly everything is for sale here. Despite its lack of appeal as a permanent residential site, the region has become a mecca for weekend escapees from the Los Angeles Basin. The formerly pristine high desert environment draws off-road enthusiasts, "rock hounds," campers, and others seeking relief from the urban grind. Even the least

JOSHUA TREE NATIONAL MONUMENT

To the east is *Joshua Tree National Monument,* which occupies over 870 square miles on the southern edge of the Mojave Desert and encompasses spectacular granite formations and thousands of Joshua trees. The massive numbers of cacti, ocotillo, palo verdes, smoke trees, and piñon pines creates a spectacularly beautiful landscape which, until recently, was a little-visited part of the desert. It offers scenic vistas of the low Colorado Desert and the Salton Sea.

observant traveler can't help but notice the enormous socioeconomic difference between Timico Acres and Palm Springs. There is truly something in the desert for everyone.

The quantity and quality of development picks up dramatically at the junction of California Highways 247 and 62 at Yucca Valley. With more than 20,000 people *Yucca Valley* serves as the gateway to the Joshua Tree National Monument and the major service center for the U.S. marine base at Twentynine Palms. Heretofore sleepy desert towns, Yucca Valley and *Morongo Valley* have experienced rapid population growth as property values in the Palm Springs area, just 20 miles away, skyrocketed. Their high desert locations also make them attractive alternatives to the lower, hotter climes of the Colorado Desert immediately to the south and east.

Dropping down more than 2,000 feet from Morongo Valley, there is a noticeable change in both temperature and social standing upon arriving in Palm Springs. Just north of the urban area a huge "wind farm," an alternative energy source, was built in the 1980s to take advantage of the locally persistent winds generated by the funnel effect created by the interface of the high desert plateau and the adjacent lowland passes. This highly inefficient source of power hardly reminds one of Dutch windmills but seemed like a good idea to people like former California governor Jerry "Moonbeam" Brown.

Palm Springs

Nestled at the base of the 10,804-foot-high Mount San Jacinto, *Palm Springs* has been a playground for the rich and famous since it was discovered by Hollywood stars in the 1920s. Located at the northern end of the Coachella Valley less than 500 feet above sea level and in the rain shadow of the San Jacinto range, its winters are warm and sunny even compared with the rest of Southern California. With its numerous exquisite golf courses and posh country clubs, Palm Springs and its satellite towns form the unquestioned elite winter resort community of California.

Originally prized by early tourists for its mineral spas and exotic palm-studded environment, the area's attractions have been diversified and increased through time. Palm Canyon Drive, Palm Springs's main thoroughfare, is lined with exclusive boutiques featuring expensive art work, jewelry, clothing, and other accoutrements of the upper classes. Renowned for its fine dining, the city boasts gourmet restaurants and chic cafés. In addition to its seasonal residents, Palm Springs boasts a thriving artist colony supported by tourists as well as local patrons.

The foothills are filled with enormous, gated mansions and hidden retreats while the flatlands to the east are filled with higher-density developments, such as condominium complexes, resort hotels, and slightly more modest single-family dwellings. Palm Springs has acquired international status and boasts a cosmopolitan population in the winter and spring that rivals any of the world's other famous playgrounds. We will take a spin down Palm Canyon Drive, the major shopping and residential street of the city, to give you a sense of the ambience of the place.

Because of its allure, beginning in the 1950s Palm Springs became a favorite haunt of college students during spring break. They were initially attracted by the hotels' relatively low rates at the beginning of the off-season, which heralded the arrival of really oppressive summer heat. Students from throughout the region flocked to the city and made Palm Springs the Fort Lauderdale of the West. Movies such as *Palm Springs Weekend* (1963) heightened the awareness of the fun to be had in the desert. However, students drunkenly cavorting through town in the nude was viewed by the city's elders as somewhat *déclassé* and subsequently discouraged by local authorities. In recent years, Palm Springs has tried to dissuade students and others of the unwashed masses from having their spring rituals on site.

The Palm Springs area offers attractions at various elevations. After a quick glimpse at the posh urban landscape, we head for the hills. Just before sunset, take the magnificent tram ride up to the 8,500-foot level of Mount San Jacinto where you will enjoy truly spectacular views of the desert spreading out below. The ride upward is exceedingly steep as the tram car narrowly avoids (we hope) the sheer granite outcrops of the mountain. Alighting from the tram, we enter an entirely different world of refreshingly cool temperatures and fragrant pine forests. Many trails lace the upper portions of the mountain, for those who want to stretch their legs or commune with nature. Dinner at the top of the tramway allows one to appreciate the unique beauty of this place.

After eating, descend from this pleasant setting in the pitch-black nothingness to your hotel in the scorching desert below.

△ Day Three

PALM SPRINGS TO SAN DIEGO, CALIFORNIA, 231 MILES

Palm Springs to Indio, California, 25 miles, Hwy. 111

Today we leave Palm Springs, the largest population center in the deserts of eastern California, and head to San Diego, which many people believe is America's finest city. But there's a lot to see along the way. During the first half of the day we make a transect through the length of the Colorado Desert from Palm Springs to the Mexican border.

The Colorado Desert occupies the Salton Trough, a triangular series of low basins between the Peninsula Ranges on the west and the Chocolate Mountains on the east and extending southward into Mexico. The trough is a structural basin covered with massive amounts of sediment. Faults crisscross the area, the most famous being the San Andreas Fault which runs along the base of the Little San Bernardino Mountains east of the Coachella. The northern part of the trough is occupied by the Coachella Valley, the southern part by the Imperial Valley, which forms the border with Mexico.

Climatically the Colorado Desert is one of the hottest places on Earth during the summer season. Temperatures in excess of 120 degrees Fahrenheit are recorded regularly in July and August, and it is common to have more than 100 consecutive days of temperatures in excess of 100 degrees. With less than three inches of annual precipitation, without irrigation the trough would be virtu-

Palm Springs, California, to San Diego, California

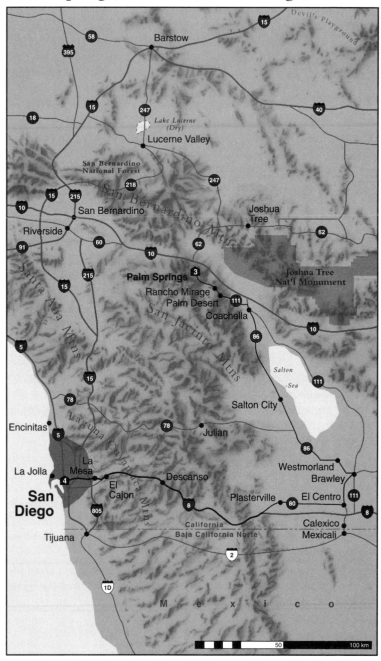

ally useless for agriculture. The flora of the Colorado Desert is similar to that of the Mojave, but is much sparser. (Although vegetation is sparse, it is still a climax vegetation.) As in the Mojave Desert, creosote bushes and burroweed are the dominant plants, but cacti and other succulents are much more common. During the brief rainy season in the spring, the desert floor comes alive in a glorious burst of color as short-lived annual flowers blossom and then rapidly fade away. In order to protect this fragile environment, the state of California has created a huge reserve in the Anza Borrego State Park and Wilderness Area.

Palm Springs sits in the northern extremity of the Coachella Valley, so over time population has grown mainly to the south where a series of communities hug the foothills of the mountains. Many of these, such as Rancho Mirage, Palm Desert, and Indian Spring, have emerged to cater to groups that, in some cases, are even more affluent than those of Palm Springs. But just 25 miles to the southeast, Highway 111 takes us to Indio and the character of the landscape starts to change markedly.

Indio to Salton City, 37 miles, Hwy. 86

Sitting astride the main east–west Phoenix–Los Angeles corridor formed by Interstate 10 and the north–south junctions of State Highways 111 and 86, *Indio*'s accessibility has given it a strong situational advantage over other towns in the Coachella Valley. An early railroad shipping point with a unique history of agricultural production, Indio wants to be known as the "Date Capital of America." The old date groves provide a picturesque and exotic landscape reminiscent of Beau Geste films.

The area was virtually an unproductive (i.e., for human use) desert until the late 1800s, but around 1900 the railroad companies developed methods of tapping artesian wells that became the basis for agriculture in the Coachella Valley. Not until 1948 did the Coachella Canal (a branch of the All-American Canal) connect farmers in the valley to Colorado River projects, which led to a boom in the region's agricultural production. One noticeable difference between the Coachella and the Imperial valleys is in farm size. In part because of labor-intensive crop types Coachella Valley farms are much smaller on average than those in the Imperial, which has led to a more densely settled farmscape. Truck crops, especially those that do well on the sandier soils in the Coachella Valley, such as grapefruits, grapes, various melons, sweet corn, and carrots, are produced and packaged here. By contrast, relatively few cattle feedlots have arisen in this valley.

Tourism is not big business in Indio, but the city is famous for its winter Date Festival which features camel races and other events not normally found in Southern California. Its economy is bolstered by Indio's status as eastern county seat for mammoth Riverside County. Indio's appearance is an attractive counterpoint to the less affluent agricultural towns of the Imperial Valley farther south.

The Salton Sea and Imperial Valley

Heading south on Highway 86, in approximately 20 miles we begin to parallel the *Salton Sea*. Occupying the lowest point in the Salton Trough, some 235 feet below sea level, the Salton Sea is a body of water some 40 miles long. The Salton Sea was formed in 1905 as a freshwater lake when flood waters from the Colorado River broke out of their levees and overwhelmed the irrigation works of the Imperial Valley. Over time the shallow (average about 20 feet deep) Salton Sea has become excessively saline because irrigation leaches salts from agricultural soils. Both the Coachella and Imperial valleys drain into the Salton Sea, thus exacerbating the flow of agricultural chemicals and fertilizers that pollute the water body.

Attempts to convert the Salton Sea into a recreational paradise have been stymied in part by its obnoxious odors and variable water level. Nevertheless, developers undertook major efforts in the 1950s and 1960s to create and market residential lots on a significant scale. Places such as Desert Shores, Salton Sea Beach,

Salton del Mar Estates, and Salton City line the northwestern shore and provide an object lesson on failed real estate schemes.

As an aside, much of the apparently serene desert environment in both the Mojave and Colorado deserts is used by the military as bombing ranges and for other ordnance-testing functions. For example, nearly all of the Chocolate Mountains east of the Salton Sea are off-limits to civilians because they are used for live strafing and bombing practice—very tough on mountain sheep.

Salton City to Brawley, California, 35 miles, Hwy. 86

The 30 miles from Salton City to Westmorland, the first of the agricultural towns in the *Imperial Valley,* could be considered desolate. *Westmorland,* however, marks the beginning of large-scale intensive irrigation agriculture in the valley. The roughly 500,000 acres of irrigated lands here represent one of the most productive agricultural regions in the United States. Agriculture here is a twentieth-century phenomenon. The irrigation network had a tenuous start in 1901 through a privately financed corporation, the California Development Company. Irregular flow of the Colorado River caused flooding and siltation—the Salton Sea is a reminder of that reality. The Colorado River Compact, signed in 1922, gave California specific water rights but not until construction of Hoover Dam (completed in 1938) was the Colorado's flow regulated to decrease the threat of disastrous flooding. About 15 miles north of Yuma on the Colorado River, the Imperial Dam was constructed as the storage reservoir for the current irrigation system, which was opened in 1940. The All-American Canal and a series of lateral canals splitting off from it carry water to farms and ranches throughout the valley. All of this effort and expenditure of federal monies was justified because of the national demand, especially in the East, for winter crops that could be grown here and shipped by rail.

The soils of the Imperial Valley are extremely fertile, the topography is flat, and the climate permits a year-round growing season, so the advent of a dependable irrigation system made this a prime site for agriculture. In addition, access to a large and growing market in Southern California and sources of low-cost labor in Mexico created a nearly ideal combination of factors for intensive agricultural production. Interestingly, the earliest farms used Japanese and Filipino laborers; not until the 1930s did Mexican workers become common in the valley. Since that time, however, the region has become increasingly a Hispanic culture area. The large scale of farming activities here and the topographic characteristics of the land are ideally suited to mechanization, thus diminishing labor requirements except in the peak harvest seasons. Specialized immigration regulations such as the Bracero Program, enacted in 1950 and abandoned in 1963, enabled landowners to bring contract laborers across the border from Mexico as needed. Currently most of the agricultural labor needs are met by U.S. citizens of Mexican origin, "green-card" holders, and an undetermined number of illegal aliens.

Westmorland is one of the once prosperous but now fading service towns that dot the valley. This *Last Picture Show* kind of place epitomizes the landscape of much of the region. With service and commercial activities increasingly being concentrated in the larger centers of Brawley and El Centro, along with a decreased farm labor population due to mechanization, the need for numerous small towns has diminished. While definitely not a boom town, Westmorland is a good place to have a cold drink and look at some of the remnant Imperial Valley architecture. Low-slung stucco buildings with arcaded façades harken back to the pre-air-conditioning era. The ubiquitous swamp coolers (which cool air by evaporating water) hang from windows and dot the rooftops, providing relief from the oppressive summer heat. Landmarks here include the central water tower with its place name prominently displayed and grain elevators with "sea-level" painted 162 feet above the ground.

Behind the shade trees, shacks, and derelict fences stand the cattle feedlots and agricultural fields. Large farm holdings domi-

Christian bikers' club, Barstow, California.

nate the valley. Despite the law that restricts federal irrigation water to farms of 160 acres or less, most units in the Imperial Valley exceed that acreage many times. Large private holdings and corporate agribusinesses constitute the bulk of landownership. The Imperial Valley is probably best known for its truck-farm production. It is among the nation's leading producers of lettuce, melons, onions, and a wide variety of other vegetables. However, most of the acreage is given over to alfalfa, sugarbeet, and cotton. Cattle are brought to the Imperial Valley from throughout the West and Mexico where they are fattened for market. Many animals, often numbering in the thousands, are concentrated in small enclosures where they eat, stand in small groups on mounds of manure, and moo the day away. The combination of heat, pungent farm produce, and manure make for an unforgettable olfactory experience. Evapotranspiration from intensely irrigated crops also creates a surprisingly high humidity and makes for a somewhat surreal desert ambience.

Just 7 miles down the road is *Brawley,* compared with most valley towns a relatively prosperous and attractive city of about 20,000. The secondary center in Imperial Valley's urban hierarchy, Brawley acts as a central place for the northern half of the region while El Centro dominates the south. Buoyed by its position astride the railroad, a number of farm implement and supply stores provide local producers with needed agricultural inputs while video rentals and grocery stores occupy much of the commercial space along the city's main thoroughfare. Brawley is experiencing some residential growth, evinced by a number of new subdivisions, such as those we'll see along Highway 86 when leaving town.

The New River, which flows through Brawley on its way to the Salton Sea, has gained a good deal of notoriety in the past several years. Strict environmental regulations in California and elsewhere in the United States have led to the establishment of noxious industries and waste-disposal facilities in Mexicali and other border cities. Unfortunately, few regulations concerning the disposal of toxic materials are enforced on the Mexican side of the border and the dumping of hazardous wastes into major waterways is common. As a result, the New River has become one of the most polluted and dangerous streams in the nation. This has prompted a good deal of ill will in addition to litigation to remedy the situation.

Brawley to Ocotillo, California, 42 miles, Hwy. 86

From Brawley we head straight south on Highway 86 to El Centro (15 miles), the major north–south arterial of the Imperial Valley. *El Centro,* as its name suggests, sits astride the major highway and rail routes serving the valley, roughly equidistant between Brawley and Mexicali to the south. The seat of government for Imperial County, El Centro has well over 30,000 inhabitants. Although the newly constructed County Building was destroyed in a major earthquake in the early 1980s, administrative functions play an important role in the economy. As the major commercial node for

MEXICALI

While there are only about 150,000 people on the U.S. side of the border in the Imperial Valley, across the border only 10 miles away the city of *Mexicali* now has nearly one million residents. Without doubt, Mexicali is the central place for Hispanics throughout the Imperial Valley and in recent years has drawn away potential shoppers from El Centro. Like its northern counterpart, the Mexican portion of the Imperial Valley is also a major center of agriculture. Because of the high salinity of water arriving from the United States, cotton production has long dominated agriculture here, but livestock and vegetable crops are gaining in importance.

Imperial Valley, El Centro sports several new shopping malls (small) and an array of other businesses. In addition to high threshold goods and services (ones that must be located in a populous area), like automobile dealers and the regional hospital, there are many motels, gas stations, and restaurants to serve highway travelers. With most of its business and service functions spreading outward, the central business district has withered and the city lacks any real "sense of place." Despite its importance to the agricultural community and proximity to Mexico, the landscape is not distinctive.

An interesting experiment in producing geothermal energy died recently with the closure of the generating plant near *Holtville* in the eastern part of Imperial Valley. The idea was to harness natural steam vents to generate power, but the operation was plagued by sediment deposits on turbine blades. This intractable problem crippled the facility's economic viability.

Leaving El Centro behind, we head westward toward San Diego. After some 10 miles of nearly flat agricultural land, we will cross the West Side Main Canal and begin ascending fairly rapidly along the arid bajadas emanating from the Laguna Mountains. *Plaster City,* linked by narrow gauge railroad to sources of raw materials in the Fish Creek Mountains to the north, is a major center of cement production for Southern California. Everything around Plaster City is covered with chalky white dust.

Some 10 miles farther west we link up with Interstate 8 at the classic desert settlement of *Ocotillo.* Once nothing more than a gas station and a garage, the town now is the southern gateway to the *Anza Borrego State Park.* A few small stores serve a growing cluster of small weekend homes for desert lovers. Periodic flash floods have wreaked havoc on the community, ripping out roads, houses, and the now-abandoned adjacent rail line.

The gradient increases dramatically west of Ocotillo as we ascend the Mountain Springs Grade where we gain 2,000 feet of elevation in only 13 miles. The eastern portion of a massive granitic batholith, locally called the Devil's Playground, is a jumble of giant reddened boulders piled in a series of spectacular formations. The next 10 miles, from Mountain Springs to Boulevard, we climb yet another 2,000 feet! This very steep grade made east–west transportation extremely difficult and thus restricted San Diego's hinterland severely. For example, producers in the Imperial Valley, less than 100 miles from San Diego, found it much less expensive to ship their goods through Los Angeles, nearly twice as distant, than to haul it over these mountains. Stop at the top of Mountain Springs Grade to appreciate the overwhelming panorama.

San Diego County

Having left the deserts below, Interstate 8 cuts its way to San Diego through the mountains via three 4,000-foot passes. Although the countryside in this area remains sparsely inhabited, we have now entered San Diego County, one of the most populous counties in the nation with more than 2.6 million residents. It will be a few hours, though, before we reach the western slope and the major urban area. First, we must get over the mountain. The sparse

JULIAN

The Sunrise Highway (S1), 20 miles from Desert View Tower at the top of Mountain Spring Grade, leads northward along the crest of the Lagunas to the quaint former gold-mining town of *Julian*. A number of campgrounds and hiking trails lace the pine forests. Cuyamaca Rancho State Park is the main recreational focus of this mountain region. Except for sunny fall weekends and periods of snow accumulation in winter, most of the facilities are relatively underutilized despite their easy proximity to the more than 2.6 million residents of the San Diego metropolitan area. The clear air and smell of the forests contrast pleasantly with the rigors of urban life.

desert vegetation on the eastern ascent, produced by the classic rain-shadow effect of the Laguna and Cuyamaca mountains which reach elevations above 6,500 feet, quickly gives way to a narrow belt of scrub juniper and chaparral. Above the 4,000-feet line in elevation, elegant pine forests provide a green canopy in marked contrast to the surrounding lower elevations. At many view points on the precipitous east slope of these mountains travelers can stand in the midst of a pine forest while looking at the desert landscape just hundreds of yards away—a textbook example of the rain-shadow effect. Precipitation at the Laguna Observatory averages nearly 40 inches annually while less than 3 inches are recorded yearly in Ocotillo.

San Diego's mountains are part of the Peninsular Ranges and part of a series of fault-block structures that slope gradually to the west and drop precipitously to the east into the Salton Trough. Much of the Cuyamaca and Laguna mountains are formed of granitic materials; the local relief is largely the result of differential weathering. With the highest peaks reaching 6,500 feet, there is a significant but not an extreme temperature gradient. As a result, precipitation is greater here than along the coastal margin and snowfall is limited during winter (however, it normally melts quickly and is insufficient to support any kind of a winter sports industry). Average winter temperatures in the higher communities, like Julian, hover around freezing while summer high temperatures average in the 80s. Coastal marine layers and the nearly constant land–sea breezes often result in summer temperatures in the mountains that are higher than on the coast. Diurnal ranges, however, are much greater.

Southern California is plagued by fires in summer and fall, and forest fires occur frequently in this part of the "back country." For example, in September 1970 a massive fire charred millions of acres from the suburbs of San Diego to the crest of the mountains along Sunrise Highway. All of the area along the interstate was engulfed in that conflagration. Fed by the huge amounts of fuel in the pine forests of the chaparral, fires often spread rapidly, aided by the hot, dry Santa Ana winds common in the early fall. While a constant threat to human occupancy, fires have done relatively little damage to property.

Many of the fire fighters in this area are of American Indian origin. They reside on the numerous Indian reservations found throughout these mountains and adjacent western valleys. Originally hunters and gatherers, the various tribal groups that lived in the region subsisted on ground acorns and local game. Living in very small groups, generally fewer than 100 persons, and scattered over a vast area, these people had relatively little impact on the physical or cultural environment. Today they represent less than one percent of the county's population but occupy a significant portion of San Diego's back country. Many of the reservations closest to the built-up urban area now operate bingo and card parlors (legalized only during the last decade), as a way of generating revenue. While hardly Las Vegas, some of the larger "casinos" attract thousands of people daily. As a marketing strategy, some even run free shuttle buses from various points in the city.

Descanso to El Cajon Valley, California, 22 miles, Hwy. 80 and I-8

To get a better idea of land use and economic activities of the area, exit at S1 the Interstate at Sunrise Highway and meander down an 8-mile loop of old Highway 80 from Pine Valley to *Descanso*. Much of the rural area, including land within the Cleveland National Forest, is used for cattle grazing. In recent years many communities, such as Pine Valley, Guatay, and Descanso, have been transformed by ex-urbanites who build "log cabins" in these settlements, partially converting them into rustic bedroom communities.

Rejoining Interstate 8 at Descanso, head the 5 miles to Alpine on the eastern fringes of the metropolitan area. During the late 1960s and early 1970s *Alpine* (about 2,000 feet above sea level) became an increasingly popular "mountain" residential community only 25 miles from the downtown San Diego. Alpine's denizens hoped to escape the air pollution and crowding of the city. Many upper-middle income families built sprawling ranch-style houses on the chaparral-covered granite knobs. Unfortunately for them, Alpine sits at the bottom of San Diego's normal inversion layer and in the path of the prevailing winds, thus making it one of the region's smoggiest locations.

A couple of miles after leaving Alpine, the large reservoir formed by El Capitan Dam can be seen on the north side of the highway. Because there are virtually no natural lakes in San Diego County, a series of dams and reservoirs provide water for drinking, irrigation, and other uses. Even the earliest Spanish settlers found it necessary to construct a dam to meet their limited needs. Throughout the late nineteenth and early twentieth centuries a number of dams were erected in the county in an effort to stabilize the water supply for a the growing urban population. Although also used for recreation, these reservoirs were made primarily to control floods and store water. Some fourteen major reservoirs were built by 1962. San Diego's reservoirs enabled it to rely entirely on local sources of water through the 1940s but since the 1960s, the region has become increasingly dependent upon water imported from the Colorado River and the California Water Aqueduct. A combination of rapid population growth, a limited number of good dam sites, and objections from conservation and environmental groups have stymied construction of additional local storage facilities.

Along this part of our route we pass through Southern California's classic vegetation assemblage—chaparral. By far the largest habitat type in the region, it is especially common from about 1,000 feet to 4,500 feet of elevation. A variety of evergreen or drought-deciduous shrubs, such as scrub oak, chamise, and manzanita, form the bulk of the vegetation. Normally growing as high as eight feet, the shrubbery when mature creates a nearly impenetrable mass. Grasses and forbs are found only in immature stands or on the margins of older assemblages, where they form an understory. The aromatic plants that comprise chaparral generate a distinctive odor closely associated with Southern California's natural environments. Despite its "native" character, chaparral stands throughout the region have been modified significantly through the introduction of exotics from the Mediterranean and elsewhere. Probably more than the higher forested areas, chaparral is prone to fire but recovers quickly from its effects. The fire-germinating characteristics of many chaparral species actually benefit from regular burning. Their extent has probably been increased due to artificial fires.

Descending into the margins of the *El Cajon Valley* on Highway 8, large mobile-park homes are frequent. As recently as the 1960s it was assumed that mobile homes would become an increasingly popular form of housing in Southern California as they have become elsewhere in the Southwest. However, the escalating cost of land during the 1970s and 1980s prevented this. Those that do exist are usually on the periphery of the urban area and their continued existence is constantly threatened even here by higher value, more intensive land uses. Because many elderly and low-income citizens live in mobile homes, legislation has been passed at both the local and state level to protect these people from eviction. It has become very difficult to close a mobile-home park without a compelling reason other than economic gain. Nevertheless, it seems unlikely that San Diego or other major Southern California cities will ever have the sprawling seas of trailers associated with places such as Phoenix, Arizona.

San Diego: Central and Eastern Residential Communities

EL CAJON TO SAN DIEGO, CALIFORNIA, 14 MILES, I-8

We first glimpse the solidly built-up San Diego area as we enter the El Cajon Valley and the suburban city of *El Cajon*. Separated from the city of San Diego by the Grossmont Summit, the relatively flat expanses of El Cajon Valley have been occupied by ranches and farms since the late nineteenth century. As late as the 1950s, the area was considered to be "small town/rural" in atmosphere and was noted for its country and western music (known in the vernacular as CW), square dances, horses, and pick-up trucks. El Cajon City was incorporated in the 1950s as a central place and service center for the surrounding farms. During the 1950s, however, the area began to grow, as widespread automobile ownership and the suburban ranch craze led many urbanites to seek space-extensive haciendas on the fringes of the San Diego region. While El Cajon was considered to be a good place to get a large house for relatively little money, low cost was sometimes associated with minimal services. Dirt roads and septic tanks were (and still are) common in many "exclusive" neighborhoods. At low densities, such an ambience could be considered rustic, but as the population has grown, there have been problems.

Over time, the social and economic diversity of El Cajon has greatly increased. With a city population of 88,692 in 1990 (and many more people living in the surrounding unincorporated areas), El Cajon has become a city in its own right. Massive "downtown" redevelopment schemes carried out during the mid-1970s led to the creation of a civic center, performing arts facility, and Parkway Plaza—a large, regional shopping center. As we travel along Interstate 8 to the junction with State Highway 67 in the heart of the valley, we see the large brown mass of the civic center complex several blocks to the south (left) and the sprawling bustle of the shopping mall immediately to the right. Also to our right as we pass Highway 67 is a large center for light industry associated with the old railroad spur to San Diego and the port and with Gillespie Field, a sizeable airport.

The hills around El Cajon Valley, especially to the south and west, now constitute some of the wealthiest residential areas in the metropolitan area. The rugged topography of neighborhoods such as Mount Helix make for a sense of privacy as well as fantastic views. In addition, increasingly strict environmental regulations make further development in such "sensitive" settings difficult if not impossible, thus ensuring that higher density and other unsuitable land uses will not intrude. Meanwhile, a great many low-cost (only by California standards), large apartment and condominium complexes have replaced tract houses in the flat "central city," thus making for a classic concentric zonal social pattern. Despite the increasing social diversity, El Cajon's population remains nearly all white. A combination of remoteness from minority communities and an image as cowboy town meant that El Cajon was virtually homogeneously white until the mid-1970s. Since then, things have changed but slowly. The city of El Cajon, for example, is still less than 3 percent black and approximately 15 percent Hispanic.

We continue on Interstate 8 over the Grossmont Summit and the small (53,000) suburb of *La Mesa*. In less than 3 miles from Parkway Plaza, we encounter Grossmont Center, another large, regional shopping center. This retail center proximity reflects, perhaps, not only a Southern California propensity for recreational shopping, but also the importance of Interstate 8 as a major transportation corridor which has "captured" the lion's share of economic activity in the eastern part of the metropolis. Indeed, this freeway is one of the ten busiest in the nation with traffic flows comparable to "big city" highways in New York City and Los Angeles. Just past Grossmont Center (at Seventieth Street) we enter the city of San Diego.

The city of San Diego, populated by more than 1.1 million people, is now the sixth largest city in the nation. Thanks to aggressive annexation policies, the city boundaries now extend from the Mexican border 15 miles south to the city of Escondido 30 miles to the north. The city encompasses more than 350 square miles and includes large amounts of undeveloped land as well as the sprawling Wild Animal Park. Thus San Diego has avoided being "strangled" by politically separate suburbs and should continue to grow for many years. San Diego has also managed to maintain socioeconomic parity with its suburban fringe so that housing quality or other dimensions of status vary little between the city and suburbs such as El Cajon and La Mesa. Many of the wealthiest areas in the metropolitan area, such as La Jolla, lie inside the city boundaries. As we enter the city at 70th Street, a neighborhood of new mini-mansions can be seen on the hills to the north (the right).

EASTERN SAN DIEGO

The city consists of a great deal more than large suburban houses, however, and we can get some idea of the diversity of the place by getting off the freeway and exploring some neighborhoods. As we exit at College Avenue and head south, the campus of San Diego State University comes into view. Although the university was founded in 1898, making it one of the oldest in California, the present campus dates from 1931. Today about 31,000 students crowd onto this compact campus.

One mile south of Interstate 8 is *El Cajon Boulevard* or old Highway 80, once the major thoroughfare to points east. This is perhaps the best example of a commercial strip in the city. The urban landscape of El Cajon Boulevard is a layered one with reminders of many past eras. There are motels, restaurants, shops, and gas stations in a variety of architectural styles from Interwar Art Deco to Postmodern Neon. We see a number of new mini-malls with yogurt and videotape-rental shops which are currently replacing the gas stations and motels of the old strip. Although the total number of businesses along the 6-mile strip has remained stable over the past fifty years, there has been nearly continuous

San Diego State University.

change in both the types of businesses and the architecture of the boulevard. Since most of the buildings have traditionally been little more than "decorated sheds," each new fantasy and whim can easily be made manifest in the commercial landscape. For example, while colorful and exuberant "Boomerang Modern" architecture prevailed during the 1950s and 1960s, many of the turrets and towers were removed during the 1970s as "good planning" and "environmental awareness" led to a greater use of wood and earth tones in harmony with the planet. Today, colorful embellishments are returning with the popularity of Postmodern architecture. While the strip has been a kind of laboratory for visual experimentation, in recent years it has also become more interesting from a social standpoint.

After traveling for about 1 mile on El Cajon Boulevard, we see the evidence of the latest social trend to affect the strip. At Euclid

Commercial strip catering to the Korean community, Artesia, California.

Avenue, we begin to see an increasing number of Asian commercial establishments. The area to the south of the strip has become one of the main residential areas for immigrants from Vietnam, Laos, and Cambodia. About 50,000 people of Indochinese origin live in the mid-city area. Many have started small restaurants, shops, and service establishments along El Cajon Boulevard. There is even an Indochinese Chamber of Commerce to organize and facilitate the expansion of Asian businesses here in mid-city. While older commercial strips in many U.S. cities have suffered abandonment and decay over the past few decades as economic activities have moved to planned suburban centers, San Diego's strips have remained viable as thousands of immigrants from Asia and Latin America have eagerly tried their luck as entrepreneurs in small business. Such businesses rely on the willingness of entire families to work long hours and to make sacrifices even more than they depend on relatively low rents in the small shops along the strip. While the fifty or so Indochinese businesses clearly dominate this part of the strip, other ethnic groups are also represented, especially Mexican and Middle Eastern.

The growth of small, immigrant businesses along the strip reveals the changing social structure of the city. While immigrant groups have traditionally clustered near the centers of U.S. cities and occupied commercial space on the fringes of downtown there is considerable evidence that this pattern has changed significantly in recent years. As urban renewal and gentrification have combined to transform many downtown areas into the elite office districts and major tourist and convention destinations, the availability of inexpensive, leftover space on the margins of the central business district has diminished. In a related development, much of the cheap housing traditionally associated with the core of the city has also been either removed or renovated and so is no longer available to newly arriving immigrant groups. Many such groups now congregate in the small apartment complexes built during the 1950s and 1960s close to old strips. The combination of housing and available commercial space in the postwar U.S. landscape has made for new types of ethnic districts. *American Graffiti* notwithstanding, much of the 1950s landscape in San Diego is now immigrant territory.

At the corner of El Cajon Boulevard and Fortieth Street, we turn to the north (right) for a look at some very different kinds of places. The intersection itself is interesting in that this section of Fortieth Street is the only remaining non-freeway portion of Interstate 15. Plans to build a freeway through the community have been discussed for over a decade and have met with much opposition. Current plans call for a highway to be constructed below ground level with a green park over it. The park would provide much needed open space in the mid-city area, and would be a focal point and give a sense of center for the community. In the meantime, however, massive clearance and abandonment associated with expectation of highway construction have led to serious social problems. Gangs and drug dealers have occupied vacant structures and have roamed through the empty lots.

Spanish architecture.

Only a few blocks north of El Cajon Boulevard, the landscape changes dramatically as we enter the neighborhood of *Kensington.* Turn right onto Monroe and then, in three blocks, left on Marlborough for a quick cruise through this serene residential area. Kensington was developed during the 1920s and 1930s as a desirable suburb with mixture of Spanish and English imagery. Most of the streets have English names while most of the houses are "Spanish" style complete with white stucco walls and red-tile roofs. This was the first community in San Diego to feature a curvilinear street pattern and underground electric wiring. Adams Avenue, once connected to downtown San Diego by a streetcar line, constitutes a small commercial district complete with a neighborhood art theater, library, and several cafes. As we travel north of Adams on Marlborough, the size of the houses increases as we approach the rim of Mission Valley. While Kensington is no longer among the most affluent communities in the San Diego area, it remains a very desirable and convenient place to live. Although many of the houses are over sixty years old, the neighborhood has never shown signs of deterioration.

Heading south again on Marlborough, turn west (right) on Adams Avenue for a look at *Normal Heights*. To do this, turn north (right) on Mountain View Drive just two blocks to the west of Highway 15. Mountain View is a twisting, meandering street that roughly follows the rim of Mission Valley. In the summer of 1985, the neighborhood was ravaged by a huge fire which started in the valley along Interstate 8 and, pushed by high winds, quickly engulfed the canyon slopes and many of the houses on the rim. The flames were capricious. Some houses were consumed while nearby houses remained unscathed. Some suffered great damage while those next door were only singed. The fire left big gaps in the urban fabric, often on lots that had the very best views over the valley. Rebuilding occurred rapidly but not without some controversy. Modest houses from the 1920s were quickly replaced by huge pink and mauve Postmodern mini-mansions. A combination of insurance money, the availability of prime view lots, and a vibrant housing market led to the construction of houses that were considerably different in scale from those nearby. Today the neighborhood offers an interesting architectural collage of San Diego house types.

Adams Avenue crosses Interstate 805; we continue for nine blocks until we get to Texas Street. This very steep and scenic roadway was one of the earliest streets to descend from the mesatop neighborhoods of San Diego into *Mission Valley*. While the mesa was developed in the years just before and after World War I, Mission Valley remained pristine agricultural land until well into the 1950s. The precipitous Mission Valley rim made for a distinctive northern edge to an early San Diego dependent upon public transportation. Streetcars could not be used on the steep grade and even early automobiles and motor buses had difficulty. With the exception of a few coastal communities, such as La Jolla, San Diego remained south of Mission Valley until the post-World War II years.

It was not only the steep cliffs that caused San Diegans to be reluctant to develop Mission Valley. The San Diego River runs through the valley and, while it is nothing more than a trickle during most of the year, it can be a raging torrent during winter storms when rainwater from a large section of the mountains comes cascading through. In rainy years, such as 1916 and 1941, the valley floor was wall to wall in water for days on end. Development began in the 1950s with the construction of the first shopping centers and hotels. With the completion of Interstate 8 through the valley in the early 1960s, development boomed. Shopping centers, convention facilities, hotels, office buildings, restaurants, and a sports stadium for professional baseball and football were completed by 1970. While the dry years of the 1960s and early 1970s made people complaisant and confident, the flooding problem had not been solved. During the 1960s, the U.S. Army Corps of Engineers developed a plan to create a concrete channel similar to those in the Los Angeles Basin, but environmentalists and others opposed the idea as ugly and ecologically inappropriate. It was soon obvious, however, that something would have to be done. A series of wet winters during the late 1970s wreaked havoc on Mission Valley developments. Roadways, parking garages, motels, and theaters were periodically underwater and traffic patterns were often disrupted.

A compromise was reached in the mid-1980s and the San Diego River Channel was dredged and banked so as to form a "green" but predictably located waterway through the valley. With flooding at least theoretically under control, new projects have proliferated over the past five years. As we descend into the valley on Interstate 8, we see large projects of many kinds—office towers, luxury hotels, shopping malls, condominiums, and recreation facilities. Turn left off Texas Street onto Friars Road and then make another left onto Mission Center Road to savor a little of this new city landscape before getting back on Interstate 8. In less than a mile, take the Highway 163 and turnoff south, toward downtown San Diego. As we head up the hill back to the mesa top, we can spend a moment pondering the ramifications of Mission Valley development on the city of San Diego. While it is true that Mission Valley has, over the years, sucked much of the economic life out of the traditional downtown (even the city's major newspapers moved there in the 1970s), the fact that the downtown and the valley are only about 3 miles apart bodes well for the economic health of the inner city. It is quite possible that the two areas will support each other in competition with more remote north and east county developments in future years.

Exiting 163 at the University Avenue exit we plunge immediately into one of San Diego's oldest and most interesting neighborhoods—*Hillcrest*. We turn right onto University Avenue and head for the heart of the district at University and Fifth. Around the turn of the twentieth century a streetcar line was built along Fifth Avenue from downtown. At University Avenue, the line split with one track heading east to mid-city and the other heading west to the exclusive area known as Mission Hills. Hillcrest, 2 miles from downtown, became the city's first trolley-generated suburban commercial node. Today Hillcrest is home to an interesting hodgepodge of life styles and social groups. A number of restaurants and coffeehouses have opened over the past decade, along with art galleries, bookstores, and an art theater. Hillcrest is the center of San Diego's gay and lesbian community as well as for a variety of urbane "dazzling urbanites." After decades of architectural stability, Hillcrest is beginning to experience growing pains. Some would argue that success is killing the community as many of the old standbys such as "The Chicken Pie Shop" are being torn down to make way for mega-malls and other projects more typical of Mission Valley. Make a loop around the core of the district and then take a look at one of the better new projects.

Heading north on Fifth Avenue, we turn east (right) onto University Avenue and continue for a few hundred yards. Soon we arrive at the uptown district, an in-fill project (more intensive land use) that makes use of an old Sears department store site. The street scene along University Avenue blends with the older buildings on the other (south) side of the street because the two-story structure has been designed to look like many small, individual buildings. The project includes apartments, condominium units, a supermarket, and several small cafes and shops. Every attempt was made to design a good project that would be "urban" rather than suburban in character and would blend into the Hillcrest ambience. We will make a loop through the area and return to the heart of Hillcrest via University Avenue. At Sixth Avenue turn left (south) for a look at the apartment district that has grown up along the west side of Balboa Park. At Laurel Street, 1 mile to the south, turn left into the park itself. We cross the Laurel Street Bridge and drive through the Prado—the cultural center of the park where the city's major museums and theaters are.

DOWNTOWN SAN DIEGO

Balboa Park (called City Park until 1915) was dedicated in 1868 when San Diego was little more than a wide place in the road. Not surprisingly, most of the park was only marginally improved over the next forty years. During the early years of the twentieth century, Kate Sessions, community activist and philanthropist, began an intensive campaign to plant trees in the park and to transform it from scrubland to something closer to the Anglo-American park ideal as epitomized by Central Park in New York City or Fairmount Park in Philadelphia. Still, the park remained largely undeveloped until the Panama–Pacific Exposition was held in San Diego in 1915–1916.

Like its counterpart in San Francisco, the San Diego Exposition was aimed at celebrating the opening of the Panama Canal and the tying together of the Americas. The architectural theme for San Diego was Spanish (more particularly from Salamanca) and it was during this era that a wide variety of "Spanish" buildings were put up not only in the park, but also in downtown and other parts of the city. Until this time, San Diego had been largely Anglo in appearance with Victorian and Romanesque structures differing little from those in the eastern states (except that they were usually made of redwood). By the time the Prado was completed, San Diego was experiencing a mania for things Spanish that, despite its ebb and flow, has not yet ended.

Today the Prado and the area around it contains the Museum of Man, the Natural History Museum, the San Diego Art Museum, the Aerospace Museum, the Old Globe Theater, the San Diego

San Diego Metro Area

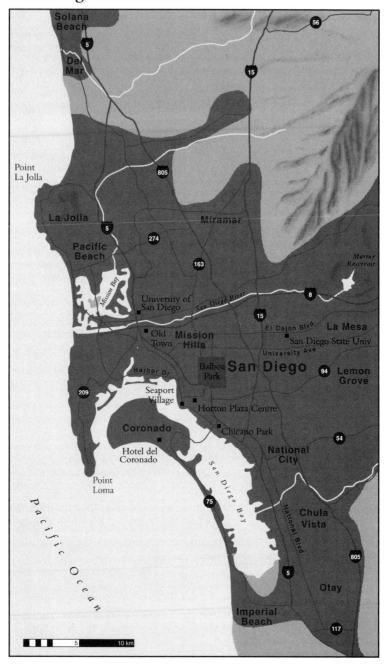

Zoo, and a wide variety of other attractions. Perhaps more than any other place, it exudes the heart and soul of the city. We exit from the Prado at Park Boulevard and head south toward downtown. Before we end our long day's journey from Palm Springs amid the hustle and bustle of center city, we take a quick look at a little more of the city's diverse physical and social environment by cruising through Golden Hill and Barrio Logan and over the Coronado Bridge to the "island" community of Coronado. First we look at Golden Hill, one of San Diego's earliest residential areas.

From Park Boulevard, we turn east (left) on Broadway and head under Interstate 5 to *Golden Hill,* passing San Diego Community College along the way. Golden Hill was developed in the 1880s and still retains much of its Victorian charm. Many Victorian mansions share the street with buildings from more recent eras including Craftsman bungalows and modern apartment complexes. Several of the more elaborate houses have been designated as official historic sites. Today the area has a mixed population, with Hispanics being most numerous. Head south on Twenty-fifth Street. The neighborhood becomes nearly homogeneously Spanish as we enter *Barrio Logan* (originally Logan Heights) south of Highway 94. We continue south for a few blocks on Twenty-fifth Street and then angle to the left on Crosby. In four blocks, we cross under the freeway (Interstate 5) and arrive at National Avenue and Chicano Park.

Chicano Park is a people's park in the best sense of the term and the symbolic center of San Diego's Hispanic community. When the Coronado Bridge was completed and connected to the Interstate freeway by a massive interchange in 1969, much of the Hispanic community that once occupied the site was decimated. More than 500 houses were destroyed and many people displaced. That the neighborhood was destroyed to create massive concrete bridge stations was bad enough, but in 1970 it was announced that a California Highway Patrol substation was to be built under the bridge as if to police what remained of the barrio. People rebelled and took over the space and demanded that a park be established. To "personalize" the brutal concrete landscape and to make it more park-like and attractive, neighborhood artists began to paint mu-

Chicano Park, under the Coronado Bridge.

rals on the pillars. Soon a program was set up by the Chicano Park Steering Committee to supervise the creation of murals. Famous artists were brought in from around the nation and from Mexico to depict the Chicano experience in California. For the past fifteen years, the park has been a center of Hispanic life in the city and many fiestas and social events are held there. The murals display everything from religious imagery to the plight of farm workers. Community members hope to expand the park so as to have a recreational shoreline along the bay like rich communities do. Speaking of rich communities, it is time to head for Coronado.

We cross the bridge to an entirely different world and get an excellent view of *San Diego Bay* and the city skyline in the process. The bridge, built in 1969, is 2 miles long, and its length as well as its graceful, curving design were necessary in order to give it sufficient height for large U.S. navy ships to pass under it. As we ascend the bridge, it is possible to observe many ships to the south as well as San Diego's small port facilities and shipbuilding yards to the north and just under the "eastern slope." While San Diego has a wonderfully protected harbor, it is not a major port; in fact it does not usually rank in the nation's top 100 in terms of tonnage. With no significant industrial or agricultural hinterland, San Diego simply relies on truck transportation to and from the ports of Long Beach and Los Angeles. The bay is thus still available for recreational use.

Our first destination in *Coronado* is the Hotel del Coronado, a Victorian masterpiece built in the 1880s which remains today as one of the largest wooden hotels in the world. Because of its unique architecture and ambience, the hotel has been used extensively in movies such as *Some Like it Hot* (1959) and *The Stunt Man* (1980). Somehow the Hotel del Coronado has kept its charm while adding amenities such as swimming pools and tennis courts. The high-rise buildings just to the south of the hotel are condominium towers built during the 1970s. Their height was controversial from the beginning and helped to create enthusiasm for the 30-foot height controls along the beach, which were enacted in the mid-1970s.

We drive along Orange Avenue through downtown Coronado to a new development on the bayside of the city known as the *Ferry*

Landing. As the name suggests, this is not only a commercial center with shops and cafes, but also a place to catch a ferry to downtown San Diego on the other side of the bay. Until the bridge was finished, Coronado could be reached only by ferry or by a long drive around the southern portion of San Diego Bay and up the strand. When the bridge was constructed, the ferry had to be discontinued until the state raised enough money from bridge tolls to pay off the bonds. When this was accomplished in the late 1980s, the ferry could operate once again and the Ferry Landing was developed to provide an attractive reason to ride. It's about a ten-minute ferry ride to downtown San Diego.

Although downtown San Diego has existed since the 1870s when the center of the tiny settlement was moved to the bay from its original location in what is now Old Town, the down-town area did not really amount to much until well into the 1980s. It started off all right with some nice Victorian office buildings and some small but interesting department stores in the 1920s. During the 1950s and 1960s, however, the downtown stagnated as new development moved to Mission Valley, Harbor Island, and a variety of suburban shopping centers and office parks. By 1970, San Diego was the largest city in the nation without a major downtown department store. Most of the big hotels and convention facilities were located either in Mission Valley or Harbor Island and so tourism bypassed the central city. In the mid-1970s, even the city's major newspapers abandoned the downtown. At that time there was only enough first-class office space in the entire core to fill the equivalent of one medium-sized skyscraper in Manhattan. There were several twenty-story glass boxes but little on the skyline that could be called architecturally interesting. Beginning in the early 1980s, however, things began to pick up.

The best way to see any compact downtown area is to walk and so we will leave our wheels behind and stretch our legs a bit. The Hotel Bar at the corner of Fifth and L (next to the railroad tracks and across from the new convention center) is as good a place to start as any. Parking is plentiful, for instance at the Horton Plaza Garage (Fourth and F).

Horton Plaza, San Diego.

The *Convention Center* on Harbor Drive at the foot of Fifth Avenue was designed to incorporate canvas "sails" over an open deck on the top floor, thus symbolically connecting the structure to the adjacent boat-filled marina. The idea was to make the center something more than just a massive box.

Downtown San Diego also boasts one of the more interesting new shopping centers in the nation. *Horton Plaza Center,* named after the historic central square on which it is located, opened in 1985 after years of discussion and redesign. Original plans called for a standard enclosed mall, but many citizens objected to the lack of connections such a complex would have with surrounding activities. The final design attempted to create an open and urban atmosphere reminiscent of an Italian city. The separate parts appear to have accreted over time rather than to have been built all at once and every attempt has been made to tie the center physically and visually with the older structures around it. Some architectural critics have suggested that the Postmodern embellishments have been overdone, but no one can deny that it is fun to experience and that it has been very successful in bringing life and retailing to downtown San Diego. To get to Horton Plaza, stroll northward on Fifth Avenue through the historic *Gaslamp District.* This was the main street of the city during the 1880s and many of the original structures remain. Today, jazz clubs and fancy restaurants coexist with street people and boxing clubs. Typical of the new urbanity is Croce's (run by musician Jim Croce's widow Ingrid) at the northwest corner of Fifth and F. We can turn west on either F or E Street to travel the one short block to Horton Plaza.

Leave Horton Plaza by the north entrance to see the historic plaza itself and to continue your stroll westward on Broadway. Broadway became the main street of San Diego during the 1920s and until the 1970s was famous for being a "sailor strip." Tattoo parlors, locker clubs, pinball arcades, bars, and stores featuring outlandish clothing dominated the scene. Today, all but a few of these activities have been eliminated. New office buildings and condominium projects now line the street. One of the most unusual is the *Emerald Center,* built to look like several individual tubes rather than one large box. Green neon lights outline the top of each cylinder making the structure a highly visible nighttime landmark.

As you approach Broadway Pier on the waterfront, you pass a large development of low-rise condominium units on your left. These opened in the early 1980s and heralded the first attempts to get people to live downtown along the magnificent harbor. They were built at suburban heights and densities so as to attract San Diegans who, by and large were unaccustomed to urban living. Today three condo projects downtown exceed thirty stories— evidence that urban living is gradually being accepted.

Broadway intersects Harbor Drive at the waterfront in an area that is gradually becoming a major tourist zone. Old vessels such as the *Star of India* (a nineteenth-century sailing ship) and the *Berkeley* (an old San Francisco ferry) as well as a number of fishing and recreational boats line the bayfront to the north. To the south, in an area that was an industrial and warehousing district until the 1970s, and still contains several industrial structures such as the Naval Supply Center, is *Seaport Village*. Seaport Village opened in the late 1970s with a combination of New England and early California architecture and a number of small shops and cafes. While it was all alone and by itself on the waterfront for nearly a decade, it proved to be quite popular and successful. Today it is part of the massive hotel and convention center complex and continues to be a popular destination for visitors. These new developments, however, have eliminated vast areas of free parking that were available in the early years, and so many locals now avoid the area. At Seaport Village we complete our circular walk with another view of the Convention Center and the new, towering condominium towers nearby.

San Diego, California, to Newport Beach, California

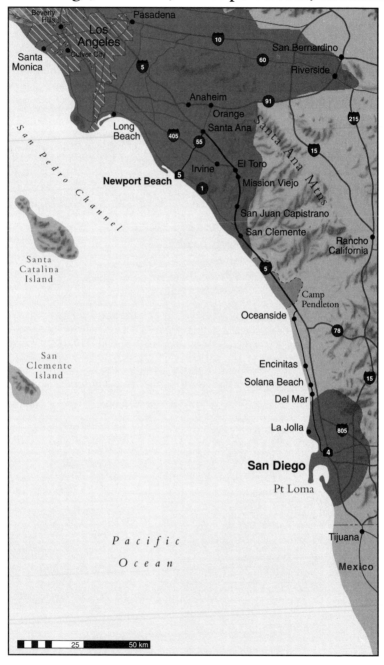

◭ Day Four

SAN DIEGO TO NEWPORT BEACH, CALIFORNIA, 83 MILES

San Diego to Del Mar, California, 13 miles

SAN DIEGO BAY

We start the morning with a trip around San Diego Bay. Heading north on Harbor Drive, in less than 2 miles from downtown we arrive at the edge of *Lindbergh Field* (so named because the famous flyer Charles Lindbergh built the *Spirit of St. Louis* here), San Diego's international airport. Having the airport in the center of the city is a mixed blessing. It is convenient to the offices and hotels of downtown—travelers can practically walk from the airport to many nearby attractions. Landing is a spectacular experience, since the planes literally fly between tall buildings when approaching the airport. On the other hand, actors in the outdoor summer theater in Balboa Park must stop mid-sentence when planes fly over at several hundred feet lest they not be heard at all. The noise and potential dangers of having a major airport in the downtown area have long been recognized. Plans to move the airport have been under discussion almost since it was established but there are no obvious alternative locations. With mountains to the east, Mexico to the south, and intensive development to the north, every location has drawbacks. Some have suggested a floating airport off the coast. Meanwhile the airport becomes busier with each passing year and there is little room for expansion.

At the airport exit, we turn off for a quick look at *Harbor Island.* Harbor Island is not really an island at all but rather an artificial peninsula or spit created in the 1960s as the bay was dredged. Today, Harbor Island is lined with large hotels and restaurants as well as public walkways. It is a popular place for walking, jogging, fishing, and just watching the sun come up or go down. It is also home to many of San Diego's recreational sailboats because of its excellent protected marina.

Continuing west on Harbor Drive, in another 2 miles we arrive at *Shelter Island,* another artificial peninsula much like Harbor Island. Shelter Island now provides a protective arm for the San Diego Yacht Club, the oldest and most venerable of San Diego's sailing establishments and current home to the America's Cup. Many boat-repair yards are here as well. At Shelter Island, Harbor Drive joins Rosecrans Street which is also Highway 209. Follow the 209 signs for a few miles to the south, through the Fort Rosecrans Military Reservation to the Cabrillo National Monument at the southern tip of Point Loma. Here we are high on a hill with views over both San Diego Bay and the Pacific Ocean. Point Loma has a visitors' center and a historic lighthouse. The old lighthouse is no longer used because it is too high to be seen easily on foggy nights. A new lighthouse has been built near sea level. For those with time and energy, a road leads down to the new lighthouse where it is fun to explore the tidepools when the sea is out.

Head northward on Highway 209 to Shelter Island and Rose-crans Street. To the right on Rosecrans is the immense Camp Nimitz, the U.S. Naval Training Center. To the left is a neighborhood of large and stately Spanish-style houses built during the 1930s. While the area is still considered to be a very attractive place to live, the attraction is somewhat mitigated by the noise of jets from the airport flying over immediately after take-off. As we enter the Midway–Rosencrans commercial area a mile or so farther along, notice the Bookstar bookstore on the right. This is an excellent example of adaptive reuse. The building was once a neighborhood theater which was restored so as to maintain its Art Deco splendor.

Passing under Interstate 5, and in less than a mile, Rosecrans becomes Taylor Street as we enter the Old Town–Presidio Park area. San Diego began at *Old Town* in 1767 and remained there for over a century until the center of activity was moved to the present downtown area on the bay in the 1870s. Although there were attempts to restore (in most cases rebuild) the best of the Spanish–Mexican structures during the early decades of the twentieth century, the Old Town area was little more than a motel strip on the road from San Diego to Los Angeles from the 1920s to the 1960s. In 1967, when San Diego celebrated its bicentennial, Old Town was refurbished and gradually turned into a major tourist attraction. Old Town State Park was established where a variety of carefully developed museums add authenticity to the experience. Today the district provides a nice mixture of heritage and fun. Mexican restaurants, cantinas, tourist shops, and museums share space in and around the park. In the past decade, large hotels have been built here also.

There have been some interesting design dilemmas in Old Town having to do with the thorny issues of preservation for authenticity versus restoration for fun as the area has been revitalized. For example, over the years the originally barren and dusty Spanish plaza (which served as a military parade ground) was planted with grass and trees so as to conform to the aesthetic tastes of Anglo settlers. Some have argued that, if the area is to be authentically restored, the plaza should be returned to it original character. So far, the trees have survived. Another dilemma revolves around what to do with structures that were built to serve highway travelers in the heart of what is now the official Old Town State Park. For example, one of the most popular eating and shopping facilities in the park is a renovated courtyard motel. It is appealing but only marginally historic. The only way to experience Old Town is to park and walk.

While you are in the area, it is a good idea to go up the hill to *Presidio Park* (either walk or drive). The San Diego Mission was here until the padres decided that it would be a good idea to move inland to the San Diego River Valley for a more reliable water supply and to get the newly converted Indian women away from

Businesses catering to the Latino market, Los Angeles.

the Spanish soldiers. During the 1920s, a park was built on the hill and *Serra Historical Museum* was established to display artifacts from early San Diego. The serenity of the park and Spanish-style museum contrasts mightily with the hum of the Interstate 8 and Interstate 5 freeways below.

Leaving Old Town, backtrack to the southeast on Taylor and Rosecrans for a few blocks and then take Sports Arena Boulevard to the right. The *San Diego Sports Arena,* for which the street is named, was built in the 1960s to attract major league basketball and hockey franchises. It has been a very limited success. San Diego has had a number of National and American Basketball Association basketball teams and several hockey teams but none of them has stayed for very long. San Diego State University basketball teams have used the arena in recent years but plans for an on-campus facility are well along. Anyway, the parking lot is used regularly for one of the largest swap meets in the city.

After a mile or so on Sports Arena, we pass under Interstate 8 and over the San Diego River into *Mission Bay Park,* one of the major playgrounds of the metropolitan area. As recently as the 1950s, Mission Bay was a rather shallow and unattractive mud flat. Although considered to be a desirable location by many of the region's birds, most people tried to have as little contact with the place as possible. In the early 1950s, the San Diego River was channelized just south of the mud flats and the old bridge that once connected Ocean Beach and Mission Beach was taken down. The history of Mission Bay as a park and recreation area begins in the early 1960s when extensive dredging and piling led to a combination of deeper water and a series of islands with regular and predictable shorelines. Since then, a variety of hotels, commercial centers, marinas, bike paths, and parks have been developed all around the bay. *Fiesta Island,* the largest of these artificial creations, is the only one that remains undeveloped. Sludge-drying operations still go on there, so it often smells pretty bad. Skirt the bay and head for Mission Beach via Mission Bay Drive.

MISSION BEACH

Mission Beach occupies a narrow strip of land to the west of Mission Bay. It was first developed as a beach community in the mid-1920s when Mission Bay Park, a large amusement park featuring a roller coaster and swimming pool (the plunge) was constructed along with a number of tiny beach cottages. The park was connected to downtown San Diego by a streetcar line. Indeed, as in many American cities, the same man (John D. Spreckels) who built the streetcar system built the park to increase ridership and encourage land speculation. A seawall mitigated but did not completely eliminate the threat of periodic flooding by storm-driven high tides, and the residential lots were gradually sold off. Mission Beach remained a rather sleepy and somewhat ramshackle area until well into the 1960s when it was known for its beach bums, hippies, and students claiming to be in their fifteenth year of pre-medical school. With the creation of Mission Bay Park as a major aquatic playground, Mission Beach became more attractive. In the 1970s, the electric utility poles along Mission Boulevard were taken down in favor of underground cables and the street was beautified and embellished. Many of the old cottages were replaced by three-story (due to the thirty-foot height limits) condo projects and shopping arcades. While there are still a few cheap rentals in the area, Mission Beach is now an expensive address. It is still a good place to sit along the seawall and watch the characters run, walk, skate, crawl, and bicycle by.

The roller coaster was closed in the mid-1970s and much of the amusement park was torn down to provide space for parking and new commercial development. After much public outrage and discussion and years of arguing over safety and public liability, the roller coaster was finally refurbished and reopened in the late 1980s. The plunge has also been rebuilt and so the former park now consists of a combination of old landmarks and new development.

As we continue northward on Mission Boulevard, Mission Beach blends into the community of Pacific Beach but the character of the street changes only slightly. Three blocks to the north at Garnett we reach Crystal Pier, one of the few remaining piers to have survived the ravages of Pacific storms. In another dozen blocks, Mission Boulevard ends. We turn left onto La Jolla Boulevard to experience yet another type of San Diego environment.

LA JOLLA

Since it is not an incorporated city but rather a prestigious neighborhood, nobody knows exactly where *La Jolla* begins and ends. We have found the density of the use of the name La Jolla is greatest on the margins of the area as everyone tries to capitalize on the fame of the community. We drive another mile or so through the neighborhood of Bird Rock before entering the old village of La Jolla. The original village of La Jolla was settled at the turn of the century as a picturesque artists' colony, remote from the center of San Diego more than 10 miles away. With the increasing use of automobiles in the 1920s and 1930s, the village expanded and elite houses were built in the hills and along the rugged coastline for their view. Still, growth in the area did not really explode until the mid-1950s when expensive tract developments began to dot the hillsides. The 1960s brought the completion of Interstate 5 to the east and the establishment of the *University of California, San Diego* to the north, and La Jolla ceased to be a village.

Although a couple of towers were built before coastal height controls, La Jolla remains a generally low-rise place. Developers have tried to cram everything possible into the community and many of the old neighborhood stores have been replaced with office buildings and art galleries.

In spite of the traffic congestion and parking problems, La Jolla really does not have the kind of problems that are associated with most urban areas. For example, a recent exciting conflict that had the community up in arms involved the opening of a McDonalds in the heart of the village. Many believed the establishment would attract the wrong kind of people to the area and allow them to experience the ambience of La Jolla on the cheap. Just past La Jolla Cove, turn left on Torrey Pines Road and head northward.

In a couple of miles we arrive at the edge of the University of California, San Diego, campus where we turn onto La Jolla Village Drive. Soon we head downhill and cross over Interstate 5 (again) and enter San Diego's "Golden Triangle." (Is there any city in the world that does not yet have a golden triangle?) Some San Diegans think this is where Los Angeles, or at least Orange County, begins. The landmarks are all massive and new. A towering Hyatt Regency Hotel in the shape of a 1930s radio dominates the gateway to the district while the University Towne Center shopping center acts as its focal point. In between are office parks, mini-malls, and seas of pastel-colored condos. This is where the action is. This is the new California. This is the Golden Triangle.

We turn left (north) onto Genesee Avenue at the University Towne Center shopping plaza and cross over Interstate 5 once more, still skirting the university campus. At North Torrey Pines Road, turn right (north) and head for Del Mar via State Route 21. On our left for the first couple of miles as we head northward, is *Torrey Pines State Reserve* and Torrey Pines Municipal Golf Course.

Soon we descend into Soledad Valley and cross over a lagoon and picturesque beach before heading up the hill to the tiny but prestigious city of Del Mar.

North County: Del Mar to San Clemente, 40 miles, Hwy. S21

Del Mar has only 4,800 people and was one of the few cities in California to actually lose population during the 1980s (–157). The combination of no new land for development, smaller households, and resistance to higher densities has meant population stability in the city proper. Del Mar, however, occupies only a very thin strip of land along the coast and all of the territory adjacent to it on the inland side belongs to the city of San Diego. San Diego recently approved new developments close to Del Mar which will have over ten times that city's population and Del Mar is concerned that it will be overwhelmed by "inlanders" flocking to its beaches. Del Mar is also the site of the first Amtrak station north of downtown San Diego and so it is becoming an increasingly popular place for north county commuters to congregate. That's what happens when you locate on a scenic stretch of coastline in a county with over two-and-a-half million people.

In only a mile or so we pass through Del Mar's little downtown and head into the little San Dieguito River Valley. Here, occupying about 25 percent of the city's land area, is the *San Diego County Fairgrounds* complete with the *Del Mar Racetrack*. As early as the 1930s and 1940s, Del Mar was a popular place for celebrities such as Bing Crosby to come and play the ponies. The attraction continues and crowds flock in during the late summer racing season.

As soon as we pass the racetrack, we enter the next of a long string of San Diego County coastal communities, *Solana Beach.* Only recently incorporated as a city, Solana Beach is also quite small in area and contains only 12,900 people. Its expansion to the east is blocked by the unincorporated but incredibly wealthy community of *Rancho Santa Fé,* a neighborhood of rambling ranches

New housing construction, typical of all of Southern California.

for movie stars and polo afficianados. Rancho Santa Fé was begun during the 1920s by the Santa Fé Railroad and continues today much as it was then. In spite of incredibly high land values, densities remain semi-rural and the landscape rustic. There is no McDonalds here.

Leaving Solana Beach, we pass through *San Elijo Lagoon* into the recently incorporated city of *Encinitas*. The string of coastal communities we are traveling through have all undergone significant changes over the past three decades. Through the 1950s, places like Solana Beach and Encinitas were sleepy but reasonably prosperous beach towns relying upon a combination of retired and resort-oriented residents, agriculture, and highway services to support the tiny populations. Most communities were unincorporated and relied on the county for minimal urban services. When the freeway (Interstate 5) was completed in the early 1960s, hurried travelers bypassed them, and the beach towns suffered serious economic problems.

Little growth or change occurred during the 1960s. Beginning with the property boom and speculative building of the 1970s, the coastal communities found themselves overwhelmed with yuppies (young, upwardly mobile professionals) escaping from the congestion of San Diego and also Los Angeles and Orange counties. Places like Encinitas seemed to be the last possible spot to get an affordable piece of the beach. By the late 1970s and early 1980s, growth was leading to a breakdown in urban services in some places and there was a rush to incorporate.

Many "towns" were too small to make a go of it on their own, however, and so some merged. The city of Encinitas was the result of such a merge and included not only the settlement by that name but also Cardiff-by-the-Sea and Leucadia. Unlike Del Mar and Solana Beach, Encinitas has been able to annex inland and has grown from 36,500 in 1980 to 55,400 in 1990. This "new city" may well become one of the larger small cities in the nation along with several other fast-growing north San Diego County municipalities such as Oceanside (128,000), Vista (72,000), and Carlsbad (63,000).

While much agricultural land has given way to residential and commercial development in recent years, one crop that seems to be competing successfully is flowers. The hills to the east of the beach towns once were ablaze with blooming flowers grown for export by air freight to the color-starved cities in colder parts of the nation. Increasingly, the flowers are grown more intensively under green sunscreens where even the benign San Diego climate can be monitored and controlled to some degree.

Fortunately, a combination of rolling topography and coastal lagoons serves to keep these fast-growing communities from growing together into one undifferentiated mass as has happened to so many smaller towns in the Los Angeles area. As we arrive at the next of these lagoons (*Bataquitos Lagoon*), we turn right onto La Costa Avenue and head for the freeway. It is time to resume our northward trek. Had we continued east, we would have quickly arrived at La Costa, a planned golf–resort community which is part

SAN LUIS REY MISSION AND SAN JUAN CAPISTRANO

Tourism in the area includes the *San Luis Rey Mission,* 4 miles inland on State Route 76. The mission, founded in 1798 as the eighteenth in the chain, has been beautifully restored and now serves as a museum. A stop either here or at the mission at *San Juan Capistrano* farther north in Orange County gives some feeling for old California.

of the city of Carlsbad. Heading north, however, we pass through the old center of that city and, after passing over two more coastal lagoons, arrive at the sprawling resort and military town of Oceanside.

Oceanside, unlike the other beach towns we have passed through, has a marina and so serves as a center for a wider variety of aquatic recreational activities. Sailing and fishing boats crowd the little harbor along with a variety of seafood restaurants and shops. There is even a highrise condo, built before regulations were passed in the 1970s restricting such buildings in the coastal zone.

Oceanside's main claim to fame revolves around its role as the downtown for *Camp Pendleton,* the sprawling, 125,000 acre U.S. marine base just to the north. For the next 18 miles after leaving Oceanside, we will travel through the military base. While tanks, soldiers, and guns and a variety of amphibious vehicles, the camp's specialty, can occasionally be seen (and heard) in Camp Pendleton, the area is usually serene and pleasant. Indeed, in spite of the occasional tromping and bombing, the area serves as an ecological preserve. San Diegans are very glad that the base is here, not so much to protect them from foreign invasion as to protect them from Los Angeles. Without Camp Pendleton, the entire region could become one giant urban blob.

After the highway checkpoint aimed at counting, if not actually stopping, part of the massive flow of undocumented visitors heading for the garment factories of Los Angeles each day, we soon arrive at *San Clemente,* the picturesque town of 41,000 people where Richard Nixon once strolled on the beach, theoretically protected by the Camp Pendleton marines. Here we leave San Diego County and enter Orange County.

Orange County: San Clemente to Newport Beach, California, 30 miles, I-5 and 405

A combination of rugged topography and environmentally conscious, elite communities has protected southern Orange County from the massive development that has occurred to the north. Indeed, Orange County's growth has slowed a bit over the past few years, allowing San Diego County to take the number two spot behind Los Angeles County. Still, Orange County had over more than 2.4 million residents in 1990 and growth is evident everywhere. The hills beside the freeway are lined with houses perched in all but the most unbuildable locations. While the hillside communities of San Clemente and San Juan Capistrano remain fairly small, the scale changes in 12 miles as we approach Mission Viejo, Laguna Hills, and the "new town" of Irvine. In the distance to our right, we can see the massive facilities of El Toro Marine Base. From here to the end of our journey in Santa Monica, the landscape is all solidly urbanized.

Just past Mission Viejo, the freeway bifurcates into Interstate 5 and Interstate 405. For years this was an agricultural area awaiting change—prime acreage awaiting development. That development is finally arriving with planned mega-malls and office parks. This nameless node is destined to be one of the major centers of activity in Southern California—yet another Golden Triangle. Just a mile or so past the split, we arrive at the city of *Irvine.* Formerly known as Irvine Ranch, a Spanish land grant, Irvine was developed as a massive new town and planned community beginning in the 1960s. It was never really a new town or even a satellite city since it is contiguous with the oldest and largest of Orange County's cities, Santa Ana. Today, however, it is a major city in its own right. With a population of more than 110,000 and a variety of commercial activities, Irvine can no longer be called a suburb. It is the home of the University of California, Irvine, campus and the John Wayne (formerly Orange County) International Airport. We exit at University Drive and head to the west (left) toward the campus. Although the area is heavily populated, the landscape seems rustic as we pass by the linear regional park that leads to the university community. Just past the campus we turn left onto State Route 73 and head into the heart of the exclusive coastal community of Newport Beach. In about 4 miles we reach Highway 1, California's coastal highway, and turn right.

Although with nearly 70,000 people, *Newport Beach* can no longer be considered a village, the fact that its central area is focused upon a series of islands, peninsulas, bays, and inlets has enabled it to avoid some of the massive developments of nearby cities and keep its beach-town charm. We skirt Newport Bay and the famous Balboa Island en route to Highway 55 where we turn left and head to the picturesque peninsula. This is a good place to relax and spend an evening watching the yachts come and go.

Newport Beach, California, to Santa Monica, California

△ Day Five

NEWPORT BEACH TO SANTA MONICA, CALIFORNIA, 50 MILES

Newport Beach to Santa Monica, California, 50 miles

Although we will not travel many miles today, there is much to see and the roadways could be congested. Our day in the greater Los Angeles area is organized around several major nodes of activities or "downtowns" along with intervening residential and industrial areas. The major nodes vary mainly in the degree to which they have been planned. For example, The Irvine/Costa Mesa node that we visit first this morning is a totally planned, new urban complex. Downtown Long Beach, on the other hand, is an old center but one that was nearly entirely rebuilt after the 1933 earthquake and during subsequent periods of enthusiastic urban renewal. Downtown Los Angeles reflects the growth patterns and architectural mix of a typical, large American city complete with everything from skyscrapers to skid row. Finally, the Century City/Westwood area is a major node carefully inserted into older but still elite residential areas and so the mix of quiet bungalows and towering office buildings is unique.

We begin by heading northeast on Highway 55 back to the junction with Interstate 405 where we go north. Here, in the vicinity of John Wayne Airport, are many of the area's mega-developments. As if to clarify the nature of the landscape, one of the towers looming over the shopping complex next to the freeway proclaims in bold letters that it is "The City." We cruise the freeway through the suburban "Plains of Id" for about 8 miles before turning west on Westminster Avenue in the "city" of that name. Here we quickly encounter the *Seal Beach Naval Weapons Station,* a large and somewhat surrealistic open space in the midst of this sea of development. As we exit the military facility, we leave Orange County and enter Los Angeles County, the most populous county in the nation with 8,863,164 people in 1990. Our first stop is the large and extremely diverse city of Long Beach.

LONG BEACH

Although it has only 430,000 people, *Long Beach* is now larger than such traditional major cities as Pittsburgh, Cincinnati, Buffalo, and Minneapolis. It has a major port, huge industrial complexes, and a wide range of ethnic diversity. It also has a wide variety of neighborhoods and housing types and quality. We first encounter the city at the neighborhood of Naples, a residential area nestled around Long Beach Marina and Alamitos Bay. As Westminster Avenue becomes Second Street, we cross over the first arm of the marina to the island which serves as the heart of the community. After a few blocks of half-million dollar "cottages," we cross onto the mainland once again. Angling to the left on Livingston, we soon reach Ocean Boulevard, the main drag of coastal Long Beach.

Although Long Beach has its share of heavy industry, smog, and ghettos, you would never know it from passing through the elite sector which runs west to east from downtown to Naples. As we drive along the edge of Bluff Park on palm-lined Ocean Boulevard, the housing density increases markedly but the quality remains high. This is the main apartment district of the city. Most of the large apartment and condos have been built along the waterfront during the past three decades, but some towering castle-like buildings date from the 1920s. In spite of its inability to expand its boundaries through annexation, Long Beach's population increased by 70,000 between 1980 and 1990. Much of the population results from increasing housing densities by replacing older houses with apartments. For apartment dwellers, a view of the water is a major attraction. To our left in *San Pedro Bay,* a number of colorful "beautified" oil rigs appear to have been built just to add a little interest to the seascape. After a few miles, we enter the compact grid of the city center.

The small blocks and relatively narrow streets of downtown Long Beach give it a distinctly urban character. Over the past two decades, however, most of the city's major economic activities have migrated to new buildings along Ocean Boulevard, leaving much of the old town underutilized if not empty. Whether the new light-rail system connecting downtown Long Beach with downtown Los Angeles and points in between will revitalize central city retailing remains to be seen. For those not in a hurry to get to Santa Monica, a ride on the light rail system through south-central Los Angeles could be an adventure. We, however, will go on to look at the waterfront and the port.

Turning left and heading over the Queens Way Bridge, we can take a closer look at the *Queen Mary.* This magnificent ocean liner was built in 1939, just in time to serve as a troop ship during World War II. After a brief period as a luxury ship during the 1950s and 1960s, she fell victim to the popularity of air travel and was converted for land duty. Today, the ship serves as hotel and convention center for the city of Long Beach. It is possible not only to tour the ship but to sleep and eat there as well. From the vicinity of the *Queen Mary,* we can view the Long Beach skyline.

It is now time to look at the other Long Beach. The port facilities, oil refineries, and heavy industries concentrated in the western portion of the city make for a very different landscape from that in the posh and serene east side. Indeed, the next 5 miles or so of our journey takes us through one of the most surrealistic and aggressively industrial landscapes in the nation. There is often total sensory involvement with the place since we can smell it and taste it as well as see it. We head north on Harbor Scenic Drive paralleling the mouth of the Los Angeles River. To our left lies the Port of Long Beach and beyond that, the Port of Los Angeles. The combined tonnage of the two ports makes them the fourth largest facility in the nation after New York, New Orleans, and Houston. The entire landscape is artificial. No protected harbor at all existed

San Pedro Bay: Los Angeles Harbor in the near distance; Long Beach Harbor in the far distance.

here before people began digging, dredging, and reshaping the land. Virtually all of the protected anchorages you see are twentieth-century creations.

We turn left (west) onto the euphemistically named Seaside Boulevard (the western extension of Ocean Boulevard) and pass by the *Long Beach Navy Shipyard* on Terminal Island. In less than 2 miles we head to the north (right) on the Terminal Island Freeway and parallel the Dominguez Channel for about 3 miles. This is the epitome of an industrial landscape with gas flames, soot, and smoke everywhere. The initial impetus for this landscape was not only the development of the port, but also the discovery of oil in this part of the Los Angeles Basin. From the early 1920s on, the area has been dotted with derricks of all shapes and sizes. The

Oil field in the city, Los Angeles Basin.

freeway ends at Willow Street and so we turn right and travel the ten blocks to the Long Beach Freeway (State Route 7) where we turn north and follow the Los Angeles River for a few miles.

SOUTH-CENTRAL LOS ANGELES

Much of the area known as south-central Los Angeles consists of working-class suburbs that grew up around the industrial corridor that extends from downtown Los Angeles to the port. Many of the communities, in fact most of them, are not part of Los Angeles at all but rather are politically independent suburbs. Today, many of these towns are predominantly black or Hispanic although there are significant numbers of whites and Asians as well. While parts of the area are rundown and numerous non-conforming land uses are mixed in among the residences, the landscape generally does not have the urban character associated with minority communities in the older industrial cities of the Northeast and Midwest. Some neighborhoods in Carson and Compton appear downright suburban with their relatively new, high-quality tract homes. Even the core of the ghetto in Watts and Florence has pleasant streets with tidy, single-family homes.

We exit the freeway as we enter *Compton*, and continue north on Long Beach Boulevard in the heart of that city. In the late 1960s, Compton became the first city in California to have a black majority. For a while it served as a middle-class suburb for blacks whose average income was higher than that of their white neighbors. Although there were signs of residential abandonment during the 1970s, population increased significantly during the 1980s to more than 90,000 people. In about 2 miles, turn left on Compton Boulevard and travel through "downtown" until we reach the railroad tracks at Alameda Avenue ten blocks to the west. Here we turn north and follow the tracks to Watts.

Watts was established in the early decades of the twentieth century, first as a farm village and later as a railroad junction on the outskirts of Los Angeles. It was never really part of the inner city but was always considered to be peripheral, even though it serves as the junction for several railroads. Many blacks moved into the area during and after World War II to work for the inter-urban railroads, in war-related industry, and later in the nearby industries of such communities as South Gate (historically a major center of auto assembly) and Vernon. Over the years, the black district along Central Avenue just south of downtown Los Angeles grew southward and merged into Watts to form the huge black district known by the 1960s simply as Watts. During the 1960s, unemployment rates soared when many railroad jobs were eliminated after the inter-urban lines closed, and Watts became a symbol of black discontent.

Unlike most hard-core ghettos, Watts has a tourist attraction. For thirty-three years, Italian immigrant tile mason Simon Rodia worked on what are now referred to as *Watts Towers*. These web-like twin towers are made of iron and covered with concrete and colorful tiles. They now form the focal point for a state park and cultural center. Turn left onto Santa Ana Boulevard and follow still more railroad tracks until we come to the corner of Wilmington and 107th. Here we experience Watts at its finest.

The isolation of Watts from surrounding communities becomes evident when you try to drive in or out of it. Most of the streets seem to end in a T-intersection or a railroad yard. With some effort, we take Wilmington to 103rd Street and then head east to Alameda where we turn north (left). Immediately turn right on Tweedy and then, in two blocks, turn left on Long Beach Boulevard—look at South Gate and Walnut Park before cruising the strip in Huntington Park. *Huntington Park,* unlike many of the communities around it, still has a vibrant and reasonably prosperous downtown. While many of the businesses have Mexican names and decorations, the business district serves as a downtown both for black areas to the west and white areas to the east. To some degree, the small shops and cafes resemble small-town California or the Los Angeles Basin in the 1950s.

Soon after leaving Huntington Park, Long Beach Boulevard curves to the left and becomes Vernon Avenue. We are now in *Vernon,* a city without people. Actually, 152 residents watch over the sprawling industrial landscape. With lots of jobs and no people-related services such as parks or schools, Vernon is famous for having the lowest property tax rate in the state.

DOWNTOWN LOS ANGELES

Leaving Vernon and crossing the tracks once again, we will travel twenty blocks or so to Figueroa Boulevard and turn right (north) to Exposition Park and the University of Southern California. The stadium at *Exposition Park* was used for both the 1932 and 1984 Olympics. It is also home for the Los Angeles Raiders and the University of Southern California Trojans. There are a number of gardens and museums in the park, but its location is considered to be a bit off-center to the cultural life of the city and so it does not play the role of Balboa Park in San Diego or Central Park in New York City as a sort of *plaza mayor* for the urban area. Just to the north of the park is the campus of the *University of Southern California.* Established in 1880, it is one of the largest and oldest

Los Angeles Metropolitan Region

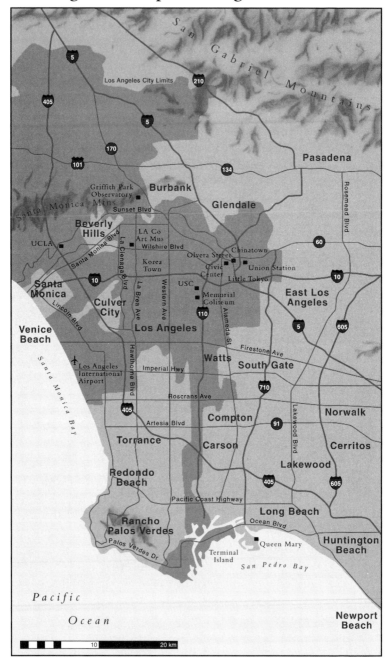

Postmodern architecture around Pershing Square, Los Angeles.

private colleges in the state. The university has played an important role in attempts to revitalize the surrounding community. It is
working with the city to encourage the movement of downtown
activities to the south so as to include the campus in the economic
heart of the metropolis. The decision to locate the massive convention center on the southern periphery of downtown was one small
step in bringing this plan to fruition. However, a big gap still exists
between the two nodes of activity.

We now enter the central core of the city of Los Angeles. With
nearly three and a half million people in 1990 (and possibly some
significant undercounting of large numbers of illegal aliens), Los
Angeles has easily passed Chicago to become the nation's second
most populous city. Compared with most American cities, Los
Angeles's landscape, population, and economic structure are remarkably diverse. The city contains a mountain range, miles of

magnificent beaches, oil fields, railroad yards, industrial complexes, skyscrapers, sprawling mansions, barrios and ghettos, movie studios, reservoirs, a major international airport, and miles of concrete river channels. The population consists of not only an incredibly diverse mix of whites, including many Armenians and Estonians, but also a great variety of Hispanics and Asians from a variety of different countries. Blacks make up about one fourth of the city's population and that group too is increasingly diverse.

The city's economy runs the gamut from high-tech computer and aerospace industries (Southern California is the largest such agglomeration in the nation) to garment and other "cheap labor" industries which often depend on illegal (and therefore powerless) workers. The older structures in and around downtown Los Angeles are a beehive of such industries. The increasingly impressive skyline of downtown Los Angeles looms large ahead of us as we proceed northward on Figueroa. We cross over the Harbor Freeway and under the Santa Monica Freeway before entering the southern fringe of downtown.

Although it is impressive from a distance and lively and colorful at street level, it is probably safe to say that downtown Los Angeles does not play the same role in the city as do the downtowns of such cities as New York City, Chicago, San Francisco, or Boston. Part of the problem is that is has never been completely clear where to put it or how big it should be. Ever since the city was founded in 1789, its center has wandered aimlessly. Because of this, the old, historic center is largely still intact and exists today as *El Pueblo de Los Angeles Historic Park,* although most people refer to it simply as *Olvera Street.* The real downtown moved south almost immediately, leaving the original plaza and the surrounding buildings as a quiet backwater. Today, many buildings from the early and mid-nineteenth century have been restored and filled with shops and cantinas. It is a good place to sit and ponder the evolution of the city. Turn right off Figueroa at Sixth Street and travel through the heart of the financial district to *Pershing Square,* the current central plaza of the city. Los Angeles had strict height limits until 1959 (nothing taller than thirteen stories except for City Hall) and so most of the skyline is new. By far the tallest

Olvera Street, Los Angeles.

structure is the recent Library Tower, which at seventy stories and 1,017 feet, is the tallest building west of Chicago. It was built using the air rights of the adjacent, Spanish-style city library. Today, Los Angeles has more than a dozen buildings which exceed 500 feet in height. It's not New York City, but then it's not Bakersfield, either. Much of the space is owned by Japanese investors.

Pershing Square, at Sixth and Olive, has been somewhat revitalized in recent years in an effort to attract a wider usage than the "street people" who have occupied its benches for decades. The Biltmore Hotel on the west side of the plaza has been restored to its former glory and adds a touch of class to the area. We will turn north (left) on Broadway two blocks past Pershing Square and drive by such local landmarks as the Bradbury Building (number 384)—a wonderfully restored office building constructed in

1893—and the Central Market—a cavernous food market with lively crowds and great neon signs. At First Street, we turn right, go two blocks to Main Street, and park in one of the underground parking garages at the civic center. It is time for a stroll and a burrito.

The southeastern portion of downtown we have just passed through is the center of the Hispanic business district, which thrives throughout the eastern part of the central business district. Most of the shops have colorful signs in Spanish, and often displays of clothing and other goods have been set up on the sidewalks. Salsa music blares from radios and loudspeakers and the smell of tortillas pervades the air. There is little to suggest that you are not in Mexico.

As we emerge from the parking garage and begin walking the two blocks northward to Olvera Street, we go to the top of City Hall for a view of the central city. *City Hall* was constructed in 1929 and, at 454 feet, was for over three decades the only skyscraper in the city. It became a famous symbol of Los Angeles when it was used for years on Sgt. Friday's badge on television's "Dragnet." This vaguely Egypto-Deco tower has an outside observation deck from which we can see the rail yards and industrial zones to the east, the civic center to the west, and the financial district to the south. Upon our descent, we stroll westward through the Civic Center Mall and the historic Olvera Street district. We may even get a glimpse of Union Station, the main railroad station which was built in 1941 just in time for the World War II hordes. Today it serves a less numerous group of Amtrak passengers.

Chinatown is just to the north of Olvera Street and *Little Tokyo* is booming just to the southeast. Return to the vehicle and mosey westward toward Santa Monica and the setting sun. There are many fascinating ways to drive to the west from downtown Los Angeles, including Sunset Strip through Hollywood and Olympic Boulevard through Koreatown. We will take the "main drag," the Wilshire Boulevard "miracle mile." The Wilshire corridor constitutes the social, economic, educational, cultural, and recreational spine of the central city. We go south on Hill Street and turn west (right) on Third where we pass through the tunnel under Bunker

City Hall, downtown Los Angeles.

Hill to the beginning of Wilshire on the west side of downtown. As we leave the west-side hotel district, we get a look at the somewhat dated, futuristic mass of the Bonaventure Hotel with its cylindrical towers and brutal base.

In a dozen blocks we arrive at *MacArthur Park,* in the heart of the west-side apartment district. The park focuses upon a pleasant lake and provides some much-needed greenery for the urban core. At Lafayette Park, four blocks farther on, Wilshire jogs to the left before beginning its straight shot to the Fairfax District and Beverly Hills. As early as the 1920s and 1930s, this stretch was emerging as a high-class commercial strip. Bullock's Department Store opened along the street and pioneered major retailing outside the downtown area.

About 2 miles past Lafayette Park, the residential area becomes quite posh, especially on the north side of Wilshire, as we enter the Country Club District at Rossmore Avenue. For a few blocks, the grid softens into an almost curvilinear pattern. In another mile we arrive in the heart of the old Jewish neighborhood of Fairfax. New York-style mid-rise apartment towers loom behind the *La Brea Tar Pits* and the *Los Angeles County Art Museum* on the north side of the street. Just to the north of the sprawling apartment complex on Third Street is Farmer's Market while Fairfax Avenue itself is lined with Jewish delicatessens and bookstores. In a few more blocks, we enter the separate municipality of Beverly Hills, one of the richest communities in the nation.

Beverly Hills consists of three types of residential areas. The southern section, where we enter the city, is a grid community with a mixture of nice houses and apartment buildings, North of downtown is a semi-grid with curving streets and mansions with large gardens. This is where most of the movie stars live. (Tourists can buy "homes of the stars" maps.) Farther north are the slopes of the Hollywood Hills–Santa Monica Mountains where a variety of lots and housing styles are tucked away. We continue on Wilshire through the commercial center of the city before turning right on Rodeo Drive, the elite, intensely "Guccied" center of Beverly Hills. In a few blocks we turn left onto Santa Monica Boulevard past the civic center. We leave Beverly Hills and re-enter Los

Venice Beach, near Los Angeles.

Angeles as we reach the swank Los Angeles Country Club and the Avenue of the Stars, gateway to Century City.

Century City lies just outside the boundaries of Beverly Hills and is a massive "new town in town" built upon land once occupied by movie studios. Century City consists of towering office blocks and shopping centers but it is also, somewhat ironically, the center of live theater in the city. For a while we seem to be back in Orange County because the new development bears little resemblance to the older neighborhoods around it. We turn south (left) on the Avenue of the Stars for a quick look. At Rancho Park, yet another country club, we turn right on Pico and three blocks later, right on Beverly Glen so we can cruise northward back to Wilshire through some typical West Los Angeles residential neighborhoods.

As we turn west on Wilshire, we enter a zone of high-rise apartments which quickly gives way to large office buildings as we approach Westwood and the campus of the University of California at Los Angeles (UCLA). The campus itself lies a few blocks to the north of Wilshire. Between the campus and Wilshire is a campus-oriented commercial district that includes a variety of restaurants, shops, and movie theaters. It is one of the few really walkable commercial areas in the metropolitan area and the sidewalks are often full, especially on weekend evenings. The campus was begun in 1929 when the university moved here from downtown. Originally considered to be a branch campus of the University of California at Berkeley, the school grew with the city and was a major center of higher education by the 1950s. We quickly loop through Westwood before continuing on to Santa Monica.

Back on Wilshire, another dozen blocks brings us to the city of *Santa Monica* on the shores of what is known as Santa Monica Bay but is really the Pacific Ocean. Santa Monica, a politically independent city of 87,000, gained a reputation for liberal politics and anti-development policies under the leadership of Tom Hayden during the 1970s. Real estate interests referred to it as "the people's republic of Santa Monica" after rent control laws were passed and limits were placed on the number and scale of new commercial projects. Compared with many of the coastal communities nearby such as Pacific Palisades and Malibu, Santa Monica has managed to keep a relatively diverse population although the wealthy young will probably eventually replace the significant population of retirees. Santa Monica and the adjacent community of *Venice* were once sleepy beach towns not dissimilar from Mission Beach in San Diego. Many of the old amusement centers, such as Ocean Park, have been destroyed but Santa Monica Pier continues and so does Muscle Beach. We turn south on Ocean Avenue and cruise the strip to Venice. If it is a sunny afternoon, the characters should be out in force.

Conclusion

Well, our tour of Southern California is over. You have had an opportunity to observe at first hand the wide variety of physical and cultural elements which make this the most dynamic region of the United States. Its ethnic diversity, the breadth of its economic base, and the range of lifestyles here are superimposed on an intriguing array of physical environments. Together, these create the unique landscapes which form Southern California and its sidekick, Las Vegas.

PART THREE

Resources

◿ Hints to the Traveler

Southern California deserves its reputation for mild weather and the corresponding casual attire. No stop on our itinerary requires specific or formal dress. Remember, however, that the route from Las Vegas to San Diego crosses a desert; daily temperatures may fluctuate from very hot at midday to cool in the evening. Cotton clothing is most comfortable. Wear long sleeves and pants, as well as a hat or visor, for protection against the intense sun. A lightweight jacket is also advisable for the evenings.

Service stations are abundant throughout the trip except the leg across the desert. Be sure that your gas tank and radiator are full before leaving Las Vegas. You may even want to pack snacks and drinking water for the long drive.

Both San Diego and Los Angeles are major metropolitan areas so adhere to the same precautions that you would in any other large city. Be aware of your surroundings. You may be perceived as an outsider in some neighborhoods and should adjust your behavior accordingly. Never be too cautious to explore; just be careful.

You do have a choice when selecting overnight accommodations with prices ranging from moderately inexpensive to outrageous. The American Automobile Association offers the best guide to medium-priced rooms. California is having a severe drought. Conserve water wherever you decide to stay. Taking short showers and not letting the water run while shaving or brushing your teeth are the most effective ways to save water.

△ Suggested Readings

Banham, Reyner. *Los Angeles: The Architecture of Four Ecologies* (Harmondsworth: Pengiun, 1973).

Banham reviews some of the ways architecture in Los Angeles might be put into environmental and/or social context. He discusses settings such as "Surfurbia," "The Foothills" and "The Plains of Id" as giving rise to architecture quite different from that found in other large American cities.

Bottles, Scott. *Los Angeles and the Automobile* (Berkeley: University of California Press, 1987).

Bottles provides an historical account of the political discussions and controversies involved in the making of the Los Angeles "Autopia." Beginning in the late 1880s, he reviews the ideas of those who would expand mass transit vs. those who would build streets and highways.

DeMarco, Gordon. *A Short History of Los Angeles* (San Francisco: Lexikos, 1987).

DeMarco provides a short, interesting, and well-illustrated account of Los Angeles from the Spanish Period to recent times all in a handy and relatively inexpensive paperback.

Garreau, Joel. *The Nine Nations of North America* (ch. 7, "Mex-America") (Boston: Houghton Mifflin, 1981) and *Edge City* (ch. 8, "Southern California"). (New York: Doubleday, 1991).

The chapters on Southern California in both books provide interesting insights into the cultural heritage and process of development in the region.

Knepp, Donn. *Las Vegas.*

This lavishly-illustrated Sunset Pictorial does a nice job of presenting the brief but lively history of Las Vegas from its early days as a railroad town and construction camp (Boulder Dam) to the rise of casinos and modern hotels.

Lantis, David; Steiner, Rodney; and Karinen, Arthur. *California: Land of Contrast* (Dubuque: Kendall/Hunt, 1977).
This is a basic regional geography book which provides lots of useful maps and basic information on everything from vegetation to climate.

Pryde, Phillip (ed.). *San Diego: An Introduction to the Region* (Dubuque: Kendall/Hunt, 1992).
This edited tome includes chapters on nearly every geographic topic imaginable from agriculture to transportation. This new edition includes updated maps and lots of data for the County as well as the City of San Diego.

Venturi, Robert; Brown, Denise Scott; and Izenour, Steven. *Learning From Las Vegas* (Cambridge: M.I.T. Press, 1972).
The authors suggest that Las Vegas represents a new and truly American kind of urban design which relates to free-standing, auto-oriented buildings rather than traditional, European-influenced ideas of vistas and street-scapes. It helps us to make sense of the seeming chaos of the Strip.

△ Index